TUNE THE HIDDEN SIGNALS ON SATELLITE TV

THE SECRET SIGNALS ON SATELLITE TV

THOMAS P. HARRINGTON W8OMV
BOB COOPER, JR. VP5D

Published By
UNIVERSAL ELECTRONICS, INC.
4555 Groves Road, Suite 3
Columbus, Ohio 43232, U.S.A.
614-866-4605

PLEASE NOTE

This work published for the commercial field to increase their awareness of the many non-video opportunities existing in the satellite communications field. The publisher and its authors do not condone any individuals, company or groups mis-use of any of this information in any way.

Section 605 of the Communications Act of 1934, as amended, generally warns you that non-broadcast-class communication signals are considered 'private' communications. The 1934 act was designed to establish guidelines to prevent unauthorized 'third parties' [i.e., anyone not associated directly with the 'private communications'] from setting out to intercept such private communications. Section 605 warns that any person not authorized to intercept such transmissions and who does so, either on purpose on accidentally, shall not divulge to any other party any of the following: (a) that such a transmission exists, (b) the content of the transmission 'intercepted'. In another section of the FCC rules, the FCC Treats al microwave [i.e., satellite] communications in the 3.7 to 4.2 GHz band as 'common carrier' transmissions, and thereafter confirms by definition that all common carrier transmissions are considered non-broadcast [as in private] transmissions. These rules were adopted at a time when the primary concern was with 'eavesdropping' on private overseas and domestic telephone communications and in studying the debate of the era it would appear that the real concern was with employees of the telco/common carrier who might have access in their work routines to these circuits. The Congress clearly was worried primarily about common carrier employees eavesdropping. In the intervening 46 years the whole nature of communications has changed including the use of some common carrier transmission frequencies for non-common carrier type signals.

ISBN #0-916661-04-0

Printed in the United States of America 1984

ACKNOWLEDGMENTS

We wish to thank the many companies and individuals who furnished us with reference materials and photos:

GTE for the cover photo in full color, The Harris Corp. — Comtech Data Corp. — Microdyne Corp. — Avcom of Virginia, Inc. — Southern Satellite Systems, Inc. — Wegener Communications Co., — Leaming Industries, Inc. — Videostar Connections, Inc. — Equatorial Communications Co. — Coop's Satellite Digest — CSD/2 — Genesis Research Corp. — R. L. Drake, Co. — Digital Electronic Systems — Arunta Engineering — Conifer Corp. — Intersat Corp. — Macom, Maco, Inc. — Paradigm Manufacturing — Sat-Tec, Inc. — United Satellite Systems — Westat Communications — KLM Electronics — Japan Radio Co. — Trio-Kenwood — Hughes Communications — Mark L. Lewis for Teletex Section — Reuters — Commodity Communications Corp. — Satellite Syndicated Systems, Inc. — Ken Prater Artwork — Mike Amirault — Mike Coyle — Bonneville Telecommunications — Scientific Atlanta, Inc. — United Video, Inc. — Dow-Jones Cable News — SSS CableText — Modulation Associates, Inc. — Winegard Company — RCA — A.T.&T. — Marshall Foiles VP-5-M

"The public's tax dollars were critical to the development of satellite communications."

Senator Barry M. Goldwater

PREFACE

Most satellites and their transponders are known by their video transmissions, however satellites also carry large amounts of non-video products that range from audio news services, news teletypewriter channels, high speed date systems, long distance telephone circuits, teletext services, stock market reports, business data channels, plus many other types of services.

Most satellite receive-only owners are aware of the many additional audio-channels available on their receivers equipped with tunable audio subcarrier sections. A popular addition to a well-equipped TVRO set-up is a quality of stereo tuner to process the stereo audio contained in the subcarrier section of the satellite signal. These signals are in addition to the regular video programs audio section, which is contained within this full subcarrier section.

In "HIDDEN SIGNALS," we will look at most of the non-video signals on the various satellites. Many of these can be received with simple, readily-obtainable equipment. Other types of services require complex computer and microprocessors to obtain the proper signal and receive intelligent copy. Most of these hi-tech services will be outlined and covered in this book.

Please be advised that the Communications Act of 1934, Section 605, which covers the rules and regulations for the interception of all private communications which are broadcast over (RF) radio frequencies, applies to interception from satellite transmissions. Contents of these transmissions are not to be divulged to anyone.

Section 605 of the Communications Act of 1934, as amended, generally warns you that non-broadcast-class communication signals are considered 'private' communications. The 1934 act was designed to establish guidelines to prevent unauthorized 'third parties' (i.e. anyone not associated directly with the 'private communications') from setting out to intercept such private communications. Section 605 warns that any person not authorized to intercept such transmissions and who does so, either on purpose or accidentally, shall not divulge to any other party any of the following: (a) that such a transmission exists, (b) the content of the transmission 'intercepted'. In another section of the FCC rules, the FCC treats all microwave (i.e. satellite) communications in the 3.7 to 4.2 GHz band as 'common carrier' transmissions, and thereafter confirms by definition that all common carrier transmissions are considered non-broadcast (as in private) transmissions. These rules were adopted at a time when the primary concern was with 'eavesdropping' on private overseas and domestic telephone communications and in studying the debate of the era it would appear that the real concern was with employees of the telco/common carrier who might have access in their work routines to these circuits. The Congress clearly was worried primarily about common carrier employees eavesdropping. In the intervening 46 years the whole nature of communications has changed, including the use of some common carrier transmission frequencies for non-common carrier type signals.

TABLE OF CONTENTS

CHAPTER ONE

UNDERSTANDING THE SATELLITE SIGNAL

The basic satellite video signal under the NTSC (National Television Standards Committee) standards for color television uses a channel band width of 6 MHz to carry all of the necessary picture information along with the audio section. This 6 MHz signal is spread over a basic 36 MHz bandwidth in order to reduce the possible terrestrial microwave interference.

The transponder spacing is at 40 MHz intervals, which allows a 2 MHz guardband between each side of the transponder or a total guard of 4 MHz.

An example is the frequency of the transponders on SATCOM F3 — Transponder 2 is 3720 MHz and 3760 on Transponder 2, or a 40 MHz spread. Of this 40 MHz bandwidth, 36 MHz is used by the video and audio requirements, plus 2 MHz either side which makes up the full 40 MHz frequency spread. This results in an actual guardband total of 4 MHz, as each transponder is afforded 2 MHz each side of its transponders. See **Figure 1.1** and **Figure 1.2.**

Most transponders have an effective bandwidth of less than 10 MHz useable frequency spectrum. This allows the transponder to carry many other signals; this open area after the required 6 MHz color bandwidth for the TV segment is where we find the many "hidden signals" on the satellite. See **Figure 1.3.**

MULTIPLE AUDIO AND DATA SUBCARRIERS

Many transponders on the satellites do not produce a video picture and therefore are considered inactive, out of service or help as a spare to replace a defective transponder. Close examination of these transponders reveal them to be active in an audio or data use.

The data use is expanded when the many narrow band audio and data carriers replace the wide-band video carriers. Further expanded use is achieved by using frequency multiplexing systems; it is possible to stack thousands of data and/or audio signals, one on top of another, in frequency order.

The total amount of non-related signals carried is dependent upon the bandwidth of the particular signals. Narrow band voice and audio channels used in most telephone circuits are approximately 4,000 Hz wide; many of these one way or single circuits are combined with many other telephone channels to form a multi-channel group, which in turn, are combined into other groups until the entire useable frequency spectrum is full.

These multiplexing schemes have been used for many years in commercial high frequency radioteletype work, and are accepted standards set by most telephone systems around the world. The International Telecommunication Union in Switzerland is responsible for setting most world standards on all types of telecommunications systems. One standard audio multiplexing system used by the major telephone systems over satellite is shown in **Figure 1.4.** This system is capable of many thousands of simultaneous satellite telephone conversations over a full-use satellite transponder. These systems in the single sideband mode, must be demodulated to be decoded, and are covered in the frequency division multiplex (FDM) section of this book.

VARIOUS TYPES OF SATELLITE TRANSMISSIONS WITH THEIR SYMBOLS

The authors have spent several years work with these "Secret Hidden Signals" on the many satellites; each of us had many volumes of data, notes, and other information collected in our search of this new satellite spectrum. As we talked and started to compare our notes and findings, we realized there was a need for a new standard to describe the various services on the satellites. This was a new part of the satellite world

Industry Standard No.	Frequency	SATCOM F3 & F4 (Polarization)	WESTAR 4 & 5 and ANIK D1 (Polarization)	COMSTAR D3 & D4 (Polarization)	GALAXY 1 (Polarization)	ANIK B (Polarization)
1	3720	1 (V)	1D (H)	1V (V)	1 (H)	1 (H)
2	3740	2 (H)	1X (V)	1H (H)	2 (V)	
3	3760	3 (V)	2D (H)	2V (V)	3 (H)	3 (H)
4	3780	4 (H)	2X (V)	2H (H)	4 (V)	
5	3800	5 (V)	3D (H)	3V (V)	5 (H)	5 (H)
6	3820	6 (H)	3X (V)	3H (H)	6 (V)	
7	3840	7 (V)	4D (H)	4V (V)	7 (H)	7 (H)
8	3860	8 (H)	4X (V)	4H (H)	8 (V)	
9	3880	9 (V)	5D (H)	5V (V)	9 (H)	9 (H)
10	3900	10 (H)	5X (V)	5H (H)	10 (V)	
11	3920	11 (V)	6D (H)	6V (V)	11 (H)	11 (H)
12	3940	12 (H)	6X (V)	6H (H)	12 (V)	
13	3960	13 (V)	7D (H)	7V (V)	13 (H)	13 (H)
14	3980	14 (H)	7X (V)	7H (H)	14 (V)	
15	4000	15 (V)	8D (H)	8V (V)	15 (H)	15 (H)
16	4020	16 (H)	8X (V)	8H (H)	16 (V)	
17	4040	17 (V)	9D (H)	9V (V)	17 (H)	17 (H)
18	4060	18 (H)	9X (V)	9H (H)	18 (V)	
19	4080	19 (V)	10D (H)	10V (V)	19 (H)	19 (H)
20	4100	20 (H)	10X (V)	10H (H)	20 (V)	
21	4120	21 (V)	11D (H)	11V (V)	21 (H)	21 (H)
22	4140	22 (H)	11X (V)	11H (H)	22 (V)	
23	4160	23 (V)	12D (H)	11V (V)	23 (H)	23 (H)
24	4180	24 (H)	12X (V)	12H (H)	24 (V)	

Figure 1.1 *Industry standard for frequency and polarization on satellites.*

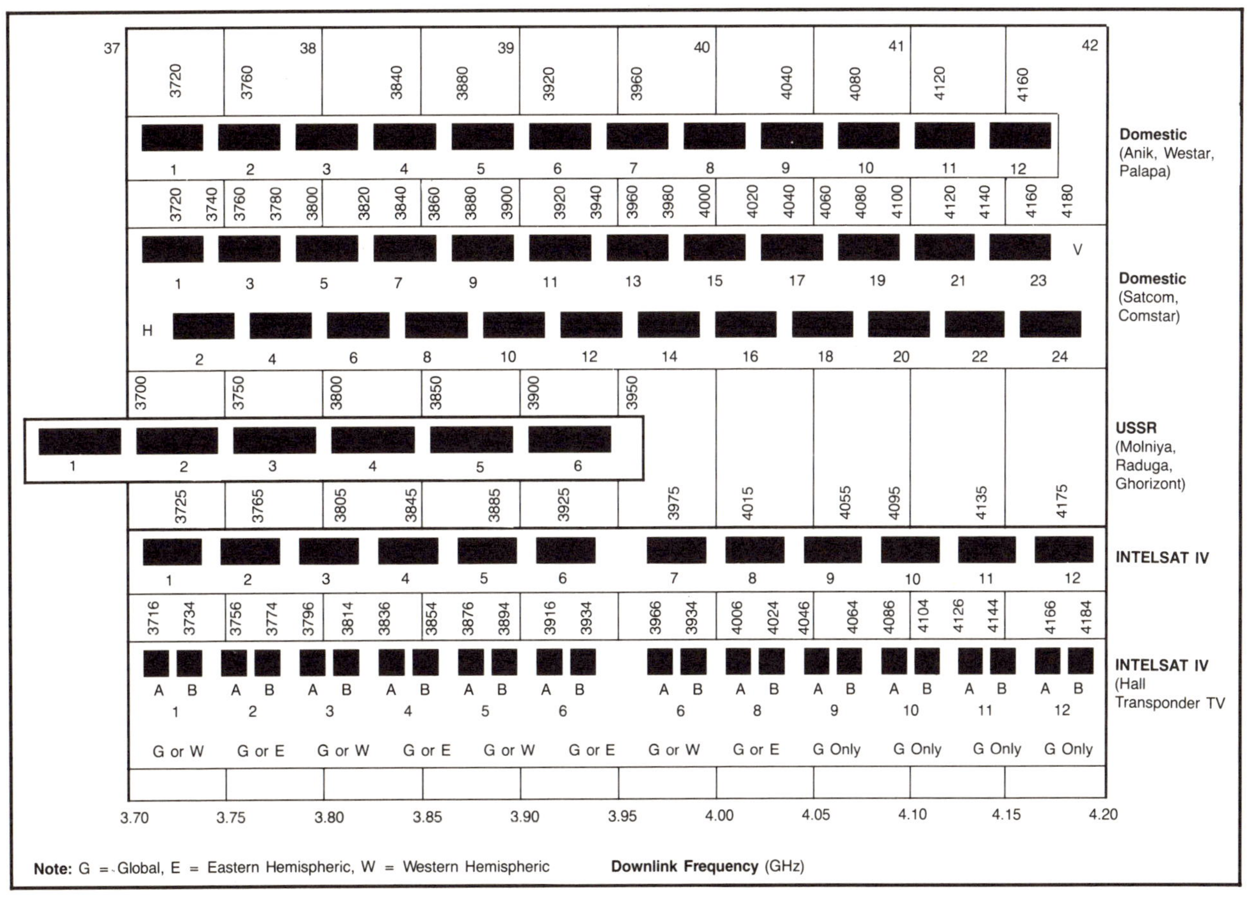

Fig. 1.2 *Transponder frequencies of the major 4 GHz satellite systems.*

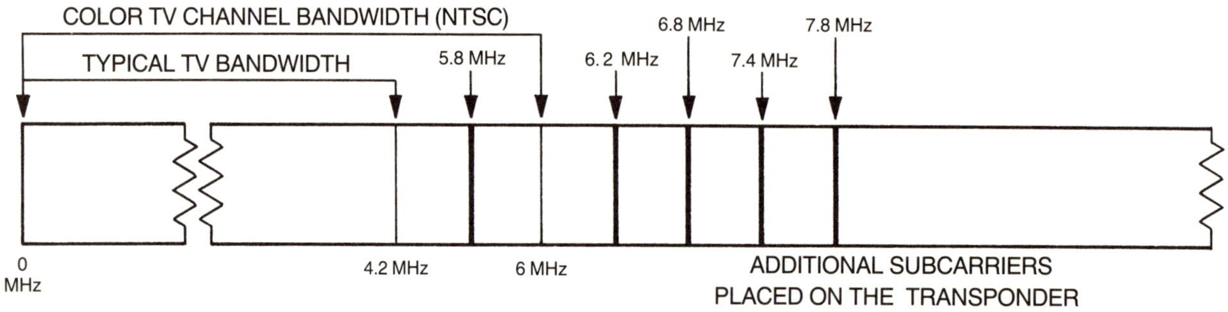

Figure 1.3 *Standard color TV carrier with audio subcarriers, program audio is on 6.2 or 6.8 MHz subcarrier.*

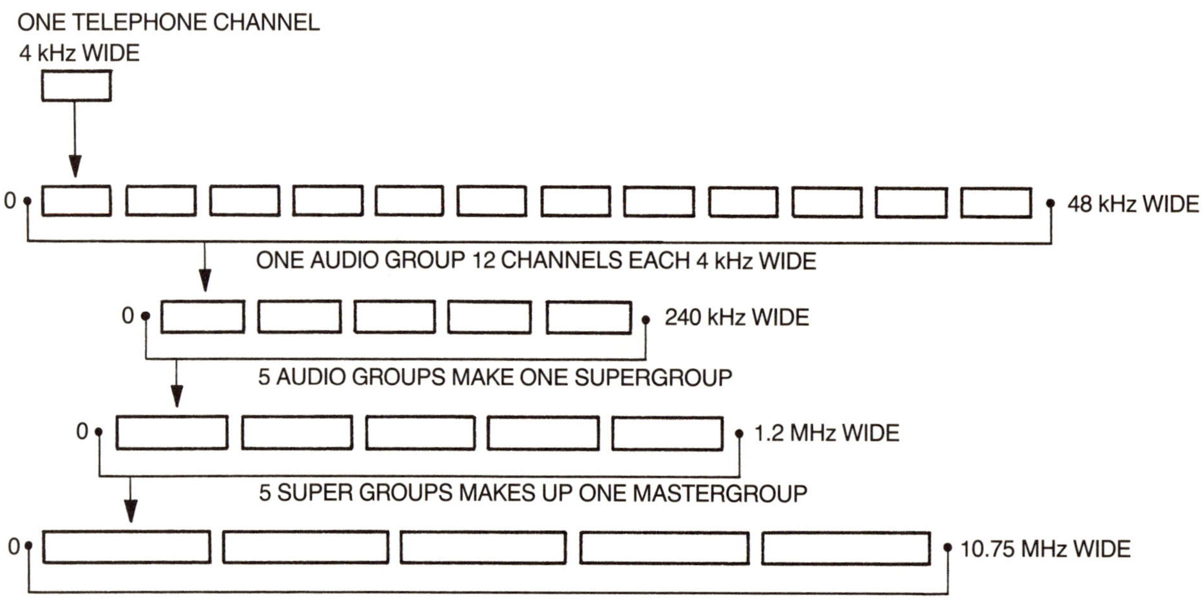

Figure 1.4 *Basic multiplexing telephone system used on many satellite systems. A single side band multiplexing mode.*

without a common language. **Figure 1.5** lists this new standard language in a simple manner, which will allow the correct compilation of information from many sources in the future.

These symbols and standards will be used in all sections of this book.

UNDERSTANDING THE SATELLITE SIGNAL — The Spectrum From 0 To 30,000 MHz

The satellite frequencies of 3.7 GHz to 4.1 GHz make up the "C" band section, along with the KU band covering 11.7 to 12.7 GHz section of the spectrum. **Table 1-A.**

The "K" band section is to be used as future direct broadcast services (DBS). The Canadian Satellite ANIK-B was the world's first dual band satellite carrying 12 C-band transponders and 6 K-band transponders; future K-band traffic load to Canada's ANIK-C series and ANIK-D series. To further understand the full spectrum of frequencies from 0 MHz to 30,000 MHz, see **Figure 1.6,** The Full Electromagnetic Spectrum.

There seems to be considerable confusion regarding the many frequency designations, Hertz—Kilohertz, Megahertz, and Gigahertz. Below is a simple table that should clear up these designations. See **Fig. 1.7.**

Major Communications Satellites Serving North America

Location: Degrees West Longitude	Satellite: Present	Number of Transponders	Future
67		24	Satcom 6** (5/86)
69		18	Spacenet 2*** (10/84)
72	Satcom 2R**	24	
74	Galaxy 2**	24	
76		24	Telstar** 302 (8/84)
76	Comstar D1/2**	24	
79	Westar 2**	12	
81		24	ASC1*** (9/85)
83	Satcom 4**	24	
86		24	Telstar 303** (5/85)
87	Comstar D3**	24	
91	Westar 3**	12	
93		24	Galaxy 3** (3/84)
96	Telstar 301**	24	
99	Westar 4**	24	
104	Anik D1**	24	
109	Anik B1***	12	
109			Anik D2** (11/84)
114	Anik A3**	12	
119	Satcom 2**	24	
123	Westar 5**	24	
127	Comstar D4**	24	
128		24	ASC2*** (9/86)
131	Satcom 3R**	24	
134	Galaxy 1**	24	
136	Satcom 1**	24	
139	Satcom 1R**	24	
143	Satcom 5**	24	
171		–	TDRS2** (late/84)

** C-Band
*** Dual C/Ku Band

Table 1–A *Communications satellites present and future.*

(A)	SINGLE SIDEBAND TYPE – DATA – FAX – MPX: SSB/SCPC
(B)	FM FORMAT TYPE – DATA – FAX – MPX: FM/SCPC
(C)	TELETEXT ON A SUBCARRIER: TxT/SC
(D)	TELETEXT ON VERTICAL BLANKING INTERVAL: TxT/VBI
(E)	DIGITAL MODULATION: DM/SCPC
(F)	SLOW SCAN TV ON VERTICAL BLANKING INTERVAL: SSTV/VBI
(G)	SPREAD SPECTRUM: SpS
(H)	SUBCARRIER USE: SbC/FM
(I)	FACSIMILE, CRT-VIDEO OR HARD COPY: FAX
(J)	MULTIPLEXING: MPX

Figure 1.5 *Language to describe the various satellite services.*

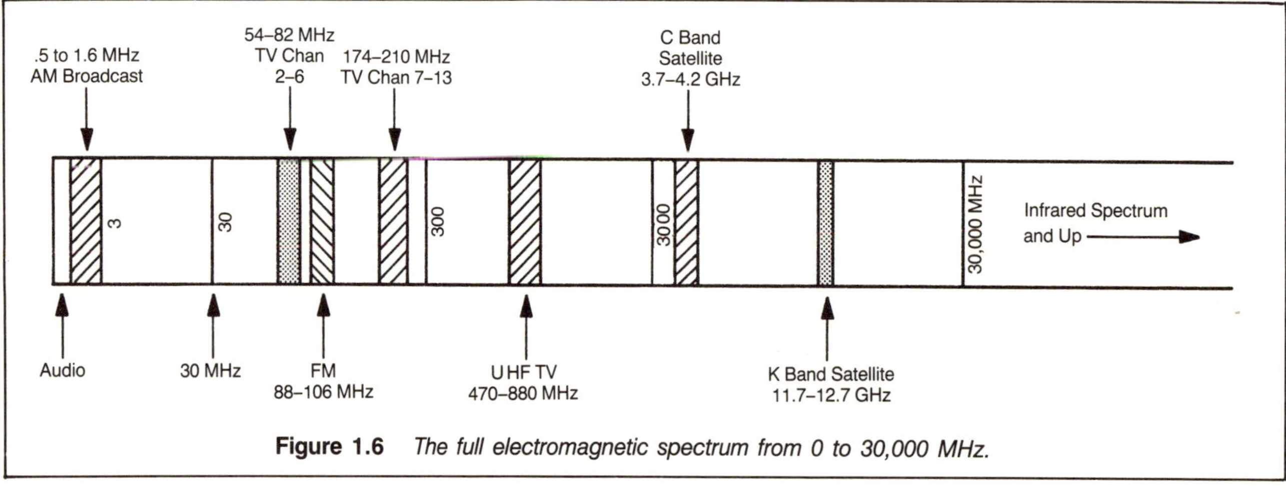

Figure 1.6 *The full electromagnetic spectrum from 0 to 30,000 MHz.*

Hz =	HERTZ =	Formerly known as cycles per second now expressed as Hertz (cycles per second) 400 Hertz (400 cycles per second)
kHz =	KILOHERTZ =	One Thousand Hertz (cycles per second) 500,000 Hz, expressed as 500 kHz.
MHz =	MEGAHERTZ =	10,000,000 Hz is expressed as 10 MHz
GHz =	GIGAHERTZ =	1,000,000,000 Hz is expressed as 10 GHz

Figure 1.7 *Frequency Designations*

BRIEF HISTORY OF U.S. SATELLITE COMMUNICATIONS

1945 Arthur C. Clarke, British science fiction writer and scientist outlined in "Wireless World" a new idea of geosynchronous satellites located in space to effect long distance communications. The method was the relay from earth to space satellite and back to earth.

1957 U.S.S.R. launches first orbital satellite, Sputnik. This starts a race for outer space with the United States.

1958 Explorer I is successfully launched and placed in orbit by the United States. Plans proceed to launch other satellites by the U.S.A.

1962 American Telephone and Telegraph Company (AT&T) launches and orbits the first active communications satellite, Echo I.

1963 The Communications Satellite Corporation (Comsat) is formed to establish a series of communications satellites to create a global satellite network.

1964 First transpacific satellite TV relayed from the Olympic games in Japan using Syncom III. Intelsat system, the international satellite consortium is established to provide multi-content satellite serviced.

1965 The "Early Bird" satellite Intelstat I successfully launched and provided the first commercial satellite pathway between the United States and Europe.

1969 Series of Intelsat III satellites are placed in orbit over the Atlantic, Pacific and Indian Oceans for worldwide commercial satellite communications.

1974 Westar I and Westar 2 launched by Western Union Company. These are the first U.S. commercial communications satellites in orbit.

1975 The start of entertainment programming via satellite begins with the first live transmission of a baseball game between the Texas Rangers and Milwaukee Brewers. The cable industry started its first satellite delivered programming with the launch of RCA's Satcom I. Also, the championship prize fight between Mohammed Ali and Joe Frazier was carried by HBO to its cable affiliates.

1976 First "superstation" via satellite is aired by Ted Turner from Atlanta, Georgia. The launch by Comsat of two communications satellites takes place, in use by AT&T and GTE. Bob Cooper, Jr. first to build and install a working home satellite TV system.

1979 RCA's Satcom F3 satellite is successfully launched and then disappears.

1980 Satellite Business Systems launched a new satellite and joins the satellite communications system users along with Comsat, RCA and Western Union. SBS will later enter the satellite telephone business.

1982 NASA's space shuttle Columbia places the first payload into space with the placement of two satellites into orbit for Telsat and Satellite Business Systems. Final approval of direct broadcast satellite systems given by The Federal Communications Commission.

1983 Successful launch and operation of Hughes Galaxy One heralds the beginning of the new higher power "C" band satellites for 24 transponders of video. The under 10 foot antenna now in wide use in many states in the U.S.

1984 Ted Turner of superstation WTBS and CNN announces a possible joint venture with HBO for a "C" band Direct Broadcast Service (DBS) using galaxy I satellites.

WHERE THE "HIDDEN SIGNALS" ARE LOCATED

This section is a "survey" of the existing domestic satellites in operation for North America. This report is a measurement tool to gauge the apparent "loading" of satellites, how busy the transponders are, and the services they carry.

The survey is done by cranking a TVRO/ARO receiving antenna toward each of the 15 domestic North American birds (Satcom 5 on the east end; including the ANIKs and then running through all the tranponders on each satellite, in both polarizations.

The basic reception equipment is a spectrum analyzer that allows the engineers to analyze and measure the content of the transponders. Video carriers, for example, have a distinctive signature pattern on a spectrum analyzer.

FDMA/FM, FDM/FM and Single Carrier Per Channel (SCPC) also have distinctive signatures. The length of the checks vary from a few minutes (ANIK A had a single transponder operation), to more than 90 minutes. The more complex the "loading scheme," the longer it takes to diagnose the type of loading present and "log" the information.

These reports miss the so-called spread spectrum transmissions (SPS). In spread spectrum, the actual frequency of the intelligence is varied up and down the transponder, or a major portion of the transponder (as much as a 2 MHz section). The receiver locks onto and follows the information as it "spreads" at a predetermined or pseudo-random rate within the assigned portion of the transponder(s). It is actually possible for spread spectrum transmissions (SPS) to fit into a transponder occupied by other more conventional modulation (carrier) formats.

Tests conducted with Intelsat birds, for example, revealed that spread spectrum data and voice could be transmitted on top of full and half transponder video, with no noticeable degradation to the video information, whereas normal "add-on" transmissions carry audio or data on subcarriers, (SbC), or on SCPC within an assigned transponder and on an assigned frequency, the spread spectrum approach literally "spreads" the transmitted information out over a big chunk of space or spectrum.

At any given frequency within the transponder, such as that occupied by the aural subcarrier for example, there is never enough, or at a level strong enough, for the user of the audio subcarrier to even notice that the spread spectrum signal is there, or was there.

On Satcom 3R Transponder 18, for example, spread spectrum shares the regular format of Reuters News Service during the daytime weekday business hours to make possible the transmission of the Reuters Small Dish Service (2 foot dish).

In the loading tables which follow, you will notice the following information under the "baseband" column:

(A) Maximum baseband 3040 kHz, tells you how much of the transponder is in use, or was in use at the time of the check. A fully "loaded" transponder would have a baseband use in excess of 8 MHz (8,000 kHz). Therefore, the lower the kHz loading number, the lower the total amount of data being transmitted on that transponder.

(B) Subcarriers: 6.2, 6.8 MHz. This tells us that in addition to the baseband "video" carrier being detected, there were also (aural) subcarriers located at the frequencies shown. It may also tell you where the center carriers are located for an FM data audio (no video) system.

(C) Subcarriers: where there are multiple audio subcarriers on such satellites as Satcom 3R, most program audio is carried on 6.8 MHz, while other subcarriers are listed as 6.2 MHz and can range from 5 MHz to as high as 8.15 MHz.

FORMAT EXPLANATION

Users should consider that some transponders noted inactive may be utilizing spot or regional beam coverage, and that emissions from these transponders may not be visible to the observing earth station.

DEFINITIONS

Baseband—The message to be transmitted or the band of frequencies transmitted. (See FDM/FM)

Digital—Wideband digital transmission of digitized voice, video, or data.

Emission—The type of radio frequency signal emitted.

Encoded—Used to indicate scrambling of the signal.

FDM/FM—A number of telephone channels are frequency division multiplexed (FDM) into a baseband which frequency modulates (FM) a carrier.

SC—Subcarrier. The total number of FM subcarriers.

SCPC—Single Channel Per Carrier. The transponder is channelized to provide a carrier for each single channel. Channels are occupied by both digital and FM carriers.

TBB—Top Baseband. The frequency of the highest baseband component.

TSC—Top Subcarrier. The frequency of the highest subcarrier.

TV/FM—A video signal (TV) and associated voice channels form a baseband which frequency modulates (FM) a carrier.

GEOSYNCHRONOUS SATELLITES

SATELLITE	WEST LONG.	TRANSPONDER/POLARIZATION SCHEME
SATCOM V	143.0	A
SATCOM IR	139.0	A
GALAXY I	134.0	B
SATCOM III	131.0	A
COMSTAR IV	127.0	A
WESTAR V	123.0	B
SPACENET I	120.0	B
ANIK A	114.0	C
ANIK B	109.0	C
ANIK D	104.5	B
WESTAR IV	99.0	B
TELSTAR 3A	96.0	A
WESTAR III	91.0	C
COMSTAR III	87.0	A
SATCOM IV	83.0	A
WESTAR II	79.0	C
COMSTAR I/II	76.0	A
GALAXY II	74.0	B
SATCOM IIR	72.0	A

TRANSPONDER
FREQUENCY/POLARIZATION

FREQUENCY	TRANSPONDER/POLARIZATION SCHEME			
	A	B	C	D
3720	1 V	1 H	1 H	
3740	2 H	2 V		1 V
3760	3 V	3 H	2 H	
3780	4 H	4 V		2 V
3800	5 V	5 H	3 H	
3820	6 H	6 V		3 V
3840	7 V	7 H	4 H	
3860	8 H	8 V		4 V
3880	9 V	9 H	5 H	
3900	10H	10V		5 V
3920	11V	11H	6 H	
3940	12H	12V		6V
3960	13V	13H	7 H	
3980	14H	14V		7 V
4000	15V	15H	8 H	
4020	16H	16V		8 V
4040	17V	17H	9 H	
4060	18H	18V		9 V
4080	19V	19H	10H	
4100	20H	20V		10V
4120	21V	21H	11H	
4140	22H	22V		11V
4160	23V	23H	12H	
4180	24H	24V		12V

SATCOM V 143.0 W

TRANSPONDER	EMISSION	BASEBAND
1	Digital – (Wideband–36 MHz)	
2	Inactive	
3	FDM/FM	TBB: 3280 KHz
4	Inactive	
5	FDMA/FDM/FM	
	3787.5 MHz	TBB: 1300 KHz
	3805.0 MHz	TBB: 2290 KHz
6	Inactive	
7	FDM/FM	TBB: 2790 KHz
8	Inactive	
9	SCPC – 20 Channels	
10	Inactive	
11	FDM/FM	TBB: 12330 KHz
12	Inactive	
13	SCPC – 4 Channels	
14	Inactive	
15	SCPC – 3 Channels	
16	Inactive	
17	FDMA/FDM/FM	TBB: 1300 KHz–1550 KHz
18	Inactive	
19	Inactive	
20	Inactive	
21	Inactive	
22	Inactive	
23	SCPC – 9 Channels	
24	Inactive	

SATCOM IR 139.0 W

TRANSPONDER	EMISSION	BASEBAND	
1	Inactive		
2	TV/FM	SC: 1	TSC: 6.8 MHz
3	Inactive		
4	FDM/FM	TBB: 9850 KHz	
5	FDM/FM	TBB: 3180 KHz	
6	Digital – (Wideband–36 MHz)		
7	FDM/FM	TBB: 8120 KHz	
8	TV/FM	SC: 1	TSC: 6.8 MHz
9	Inactive		
10	Inactive		
11	Inactive		
12	TV/FM	SC: 2	TSC: 6.8 MHz
13	TV/FM	SC: 1	TSC: 7.2 MHz
14	FDM/FM	TBB: 3740 KHz	

15	Inactive		
16	Inactive		
17	TV/FM	SC: 2	TSC: 6.8 MHz
18	Digital – (Wideband–36 MHz)		
19	Digital – (Wideband–36 MHz)		
20	TV/FM	SC: 2	TSC: 6.8 MHz
21	FDM/FM	TBB: 10340 KHz	
22	Inactive		
23	Digital – (Wideband–36 MHz)		
24	Digital – (Wideband–36 MHz)		

GALAXY I 134.0 W

TRANSPONDER	EMISSION	BASEBAND	
1	TV/FM	SC: 1	TSC: 6.2 MHz
2	TV/FM	SC: 4	TSC: 6.8 MHz
3	TV/FM (Encoded)	SC: None	
4	TV/FM	SC: 2	TSC: 6.8 MHz
5	Inactive		
6	TV/FM	SC: 2	TSC: 6.8 MHz
7	TV/FM	SC: 3	TSC: 7.2 MHz
8	TV/FM	SC: 2	TSC: 7.2 MHz
9	Inactive		
10	TV/FM (Encoded)	SC: None	
11	TV/FM	SC: 1	TSC: 6.8 MHz
12	TV/FM	SC: 2	TSC: 6.8 MHz
13	TV/FM	SC: 1	TSC: 6.8 MHz
14	TV/FM	SC: 2	TSC: 6.8 MHz
15	TV/FM	SC: 5	TSC: 7.6 MHz
16	Inactive		
17	TV/FM	SC: 2	TSC: 6.8 MHz
18	Inactive		
19	TV/FM	SC: 1	TSC: 6.8 MHz
20	TV/FM	SC: 2	TSC: 6.8 MHz
21	TV/FM	SC: 2	TSC: 6.8 MHz
22	Inactive		
23	TV/FM	SC: 1	TSC: 6.8 MHz
24	TV/FM	SC: 1	TSC: 6.8 MHz

SATCOM III 131.0 W

TRANSPONDER	EMISSION	BASEBAND	
1	TV/FM	SC: 1	TSC: 6.8 MHz
2	TV/FM	SC: 5	TSC: 7.5 MHz
3	TV/FM	SC: 21	TSC: 8.5 MHz
4	TV/FM	SC: 2	TSC: 6.8 MHz
5	TV/FM	SC: 2	TSC: 6.8 MHz

6	TV/FM	SC: 10	TSC: 7.8 MHz
7	TV/FM	SC: 4	TSC: 6.8 MHz
8	TV/FM	SC: 7	TSC: 6.8 MHz
9	TV/FM	SC: 2	TSC: 6.8 MHz
10	TV/FM	SC: 2	TSC: 6.8 MHz
11	TV/FM	SC: 3	TSC: 7.0 MHz
12	TV/FM	SC: 1	TSC: 6.8 MHz
13	TV/FM	SC: 2	TSC: 6.8 MHz
14	TV/FM	SC: 3	TSC: 7.2 MHz
15	TV/FM	SC: 2	TSC: 7.2 MHz
16	TV/FM	SC: 2	TSC: 6.8 MHz
17	TV/FM	SC: 2	TSC: 6.8 MHz
18	TV/FM (Encoded)	SC: None	Reuters Service
19	TV/FM	SC: 1	TSC: 6.8 MHz
20	TV/FM	SC: 1	TSC: 6.8 MHz
21	TV/FM	SC: 1	TSC: 6.8 MHz
22	TV/FM	SC: 2	TSC: 6.8 MHz
23	TV/FM	SC: 1	TSC: 6.8 MHz
24	TV/FM	SC: 2	TSC: 6.8 MHz

SPACENET I 120.0 W

TRANSPONDER	EMISSION	BASEBAND	
3	TV/FM	SC: 2	TSC: 6.8 MHz
11	TV/FM	SC: 2	TSC: 6.8 MHz

ANIK A 114.0 W

TRANSPONDER	EMISSION	BASEBAND
12	SCPC – 24 Channels	

ALL OTHER TRANSPONDERS INACTIVE

ANIK B 109.0 W

TRANSPONDER	EMISSION	BASEBAND	
1	SCPC – 17 Channels		
2	SCPC – 54 Channels		
3	SCPC – 6 Channels		
4	TV/FM *1	SC: 1	TSC: 6.8 MHz
5	Inactive		
6	TV/FM *1	SC: 1	TSC: 6.8 MHz
7	TV/FM	SC: 1	TSC: 6.8 MHz
8	TV/FM *1	SC: 1	TSC: 6.8 MHz
9	TV/FM *1	SC: 1	TSC: 6.8 MHz
10	TV/FM *2	SC: 1	TSC: 6.8 MHz
11	Inactive		
12	Inactive		

COMSTAR IV 127.0 W

TRANSPONDER	EMISSION	BASEBAND	
1	FDM/FM	TBB: 8480 KHz	
2	Inactive		
3	FDM/FM	TBB: 8480 KHz	
4	FDM/FM	TBB: 5770 KHz	
5	TV/FM (Encoded)	SC: 2	TSC: 6.8 MHz
6	Inactive		
7	FDM/FM	TBB: 7280 KHz	
8	Inactive		
9	TV/FM (Encoded)	SC: 2	TSC: 6.8 MHz
10	Inactive		
11	TV/FM (Encoded)	SC: 2	TSC: 6.8 MHz
12	Inactive		
13	TV/FM	SC: 2	TSC: 6.8 MHz
14	Inactive		
15	Unidentifiable – (Low C/N: 1.5 dB in 25 MHz		
16	FDM/FM	TBB: 7220 KHz	
17	Inactive		
18	TV/FM	SC: 4	TSC: 6.8 MHz
19	Inactive		
20	Inactive		
21	FDM/FM	TBB: 8520 KHz	
22	Inactive		
23	FDM/FM	TBB: 8520 KHz	
24	Unidentifiable – (Low C/N)		

WESTAR V 123.0 W

TRANSPONDER	EMISSION	BASEBAND	
1	TV/FM	SC: 3	TSC: 6.8 MHz
2	TV/FM	SC: 2	TSC: 6.8 MHz
3			
4	TV/FM	SC: 1	TSC: 6.8 MHz
5	TV/FM	SC: 2	TSC: 8.25 MHz
6	SCPC – 4 Channels		
7	TV/FM	SC: 2	TSC: 6.8 MHz
8	TV/FM	SC: 4	TSC: 6.8 MHz
9	SCPC – 22 Channels		
10	TV/FM	SC: 2	TSC: 6.8 MHz
11	TV/FM	SC: 3	TSC: 6.8 MHz
12	TV/FM	SC: 2	TSC: 6.8 MHz
13	SCPC – 21 Channels		
14	TV/FM	SC: 2	TSC: 6.8 MHz
15	TV/FM	SC: 1	TSC: 6.8 MHz

16	TV/FM	SC: 2	TSC: 6.8 MHz
17	TV/FM	SC: 4	TSC: 6.8 MHz
18	Inactive		
19	Digital – (Wideband–36 MHz)		
20	Inactive		
21	TV/FM	SC: 2	TSC: 6.8 MHz
22	TV/FM	SC: 1	TSC: 6.8 MHz
23	TV/FM	SC: 4	TSC: 6.8 MHz
24	TV/FM	SC: 2	TSC: 6.8 MHz

ANIK D 104.5 W

TRANSPONDER	EMISSION	BASEBAND	
1	Inactive		
2	Inactive		
3	Digital – (Wideband–36 MHz)		
4	Inactive		
5	Inactive		
6	Inactive		
7	Inactive		
8	TV/FM (Encoded)	SC: 4	TSC: 7.6 MHz
9	TV/FM (Encoded)	SC: 3	TSC: 6.8 MHz
10	Inactive		
11	Inactive		
12	Inactive		
13	TV/FM	SC: 2	TSC: 6.8 MHz
14	TV/FM (Encoded)	SC: 3	TSC: 6.8 MHz
15	Inactive		
16	TV/FM	SC: 5	TSC: 6.8 MHz
17	Digital – (Wideband–36 MHz)		
18	TV/FM (Encoded)	SC: 3	TSC: 6.8 MHz
19	Inactive		
20	TV/FM	SC: 2	TSC: 6.8 MHz
21	TV/FM	SC: 2	TSC: 6.8 MHz
22	TV/FM (Encoded)	SC: 4	TSC: 7.5 MHz
23	TV/FM (Encoded)	SC: 3	TSC: 6.8 MHz
24	TV/FM	SC: 6	TSC: 6.8 MHz

WESTAR IV 99.0 W

TRANSPONDER	EMISSION	BASEBAND	
1	SCPC – 12 Channels		
2	SCPC – 21 Channels		
3	SCPC – 31 Channels		
4	Inactive		
5	Inactive		
6	TV/FM	SC: 2	TSC: 6.8 MHz
7	SCPC – 5 Channels		

8	SCPC – 7 Channels		
9	SCPC – 20 Channels		
10	Digital – (Wideband–14 MHz)		
11	TV/FM	SC: 3	TSC: 6.8 MHz
12	SCPC – 15 Channels		
13	SCPC – 12 Channels		
14	FDMA/FDM/FM		
	3874.0 MHz	TBB: 3030 KHz	
	3993.0 MHz	TBB: 1790 KHz	
15	TV/FM	SC: 3	TSC: 6.8 MHz
16	TV/FM	SC: 2	TSC: 6.8 MHz
17	TV/FM	SC: 4	TSC: 7.9 MHz
18	TV/FM (Encoded)		
19	TV/FM	SC: 2	TSC: 7.4 MHz
20	Inactive		
21	TV/FM	SC: 2	TSC: 6.8 MHz
22	SCPC – 2 Channels		
23	TV/FM	SC: 2	TSC: 6.8 MHz
24	FDMA/FDM/FM		
	4172.0 MHz	TBB: 2790 KHz	
	4190.0 MHz	TBB: 2040 KHz	

TELSTAR 3A 96.0 W

TRANSPONDER	EMISSION	BASEBAND	
1	Inactive		
2	TV/FM	SC: 2	TSC: 6.4 MHz
3	TV/FM	SC: 2	TSC: 6.8 MHz
4	Inactive		
5	SCPC – 3 Channels		
6	TV/FM	SC: 2	TSC: 6.8 MHz
7	Inactive		
8	FDM/FM	TBB: 7240 KHz	
9	TV/FM	SC: 8	TSC: 7.5 MHz
10	TV/FM	SC: 2	TSC: 6.8 MHz
11	TV/FM	SC: 2	TSC: 6.8 MHz
12	TV/FM	SC: 2	TSC: 6.8 MHz
13	TV/FM	SC: 2	TSC: 6.8 MHz
14	Carrier – Unmodulated		
15	TV/FM	SC: 1	TSC: 5.8 MHz
16	Inactive		
17	Inactive		
18	Inactive		
19	Carrier – Subcarrier of 9.6 MHz		
20	Inactive		
21	Inactive		
22	Inactive		
23	TV/FM	SC: 2	TSC: 6.8 MHz
24	Inactive		

WESTAR III 91.0 W

TRANSPONDER	EMISSION	BASEBAND	
1	SCPC – 10 Channels		
2	SCPC – 26 Channels		
3	TV/FM	SC: 1	TSC: 6.2 MHz
4	SCPC – 29 Channels		
5	SCPC – 12 Channels		
6	Digital – (Wideband–36 MHz)		
7	SCPC – 29 Channels		
8	Digital – (Wideband–36 MHz)		
9	Digital – (Wideband–36 MHz)		
10	Inactive		
11	Inactive		
12	TV/FM	SC: 2	TSC: 6.8 MHz

COMSTAR III 87.0 W

TRANSPONDER	EMISSION	BASEBAND	
1	Inactive		
2	Inactive		
3	SCPC – 17 Channels		
4	SCPC – 16 Channels		
5	SCPC – 9 Channels		
6	Inactive		
7	Inactive		
8	Inactive		
9	Inactive		
10	TV/FM	SC: 2	TSC: 6.4 MHz
11	Inactive		
12	SCPC – 9 Channels		
13	SCPS – 3 Channels		
14	Inactive		
15	Inactive		
16	Inactive		
17	TV/FM	SC: 2	TSC: 6.4 MHz
18	Inactive		
19	Inactive		
20	Inactive		
21	Inactive		
22	Inactive		
23	Inactive		
24	Inactive		

SATCOM IV 83.0 W

TRANSPONDER	EMISSION	BASEBAND	
1	Inactive		
2	TV/FM	SC: 2	TSC: 6.8 MHz
3	TV/FM	SC: 11	TSC: 7.75 MHz
4	TV/FM	SC: 1	TSC: 6.8 MHz
5	Inactive		
6	Inactive		
7	TV/FM	SC: 15	TSC: 8.15 MHz
8	TV/FM	SC: 1	TSC: 6.8 MHz
9	Inactive		
10	Inactive		
11	TV/FM	SC: 1	TSC: 6.8 MHz
12	TV/FM	SC: 1	TSC: 6.8 MHz
13	Inactive		
14	TV/FM	SC: 2	TSC: 6.8 MHz
15	TV/FM	SC: 2	TSC: 6.8 MHz
16	Inactive		
17	TV/FM	SC: 4	TSC: 6.8 MHz
18	Inactive		
19	TV/FM	SC: 1	TSC: 6.8 MHz
20	TV/FM	SC: 2	TSC: 6.8 MHz
21	TV/FM	SC: 2	TSC: 6.8 MHz
22	TV/FM	SC: 2	TSC: 6.8 MHz
23	TV/FM	SC: 2	TSC: 6.8 MHz
24	TV/FM	SC: 1	TSC: 6.8 MHz

WESTAR II 79.0 W

TRANSPONDER	EMISSION	BASEBAND
1	FDMA/FDM/FM	
	3711.0 MHz	TBB: 1300 KHz
	3729.0 MHz	TBB: 2040 KHz
2	Inactive	
3	Inactive	
4	FDM/FM	TBB: 5260 KHz
5	FDM/FM	TBB: 6500 KHz
6	Inactive	
7	Inactive	
8	FDM/FM	TBB: 6500 KHz
9	FDMA/FDM/FM	
	4030.5 MHz	TBB: 2460 KHz
	4048.5 MHz	TBB: 2460 KHz
10	Inactive	
11	Inactive	
12	Inactive	

COMSTAR I/II 76.0 W

TRANSPONDER	EMISSION	BASEBAND
1	FDM/FM	TBB: 4490 KHz
2	Inactive	
3	FDM/FM	TBB: 5730 KHz
4	FDM/FM	TBB: 6880 KHz
5	Inactive	
6	FDM/FM	TBB: 5730 KHz
7	FDM/FM	TBB: 7240 KHz
8	Inactive	
9	Inactive	
10	Inactive	
11	Inactive	
12	Inactive	
13	FDM/FM	TBB: 8440 KHz
14	FDM/FM	TBB: 5480 KHz
15	FDM/FM	TBB: 7240 KHz
16	FDM/FM – (Low C/N)	
17	FDM/FM	TBB: 8460 KHz
18	Inactive	
19	FDM/FM	TBB: 5480 KHz
20	Inactive	
21	FDM/FM	TBB: 4490 KHz
22	FDM/FM	TBB: 6870 KHz
23	Inactive	
24	Inactive	

GALAXY II 74.0 W

TRANSPONDER	EMISSION	BASEBAND
6	SCPC – 2 Channels	
21	TV/FM	SC: 2 TSC: 6.8 MHz
23	Unmodulated Carrier – Dispersed 2 MHz	
24	Unmodulated Carrier – Dispersed 1.8 MHz	

ALL OTHER TRANSPONDERS INACTIVE

CHAPTER TWO

AUDIO SUBCARRIERS (SbC/FM)

AUDIO SUBCARRIERS ON SATELLITE TV

The video on satellite TV is frequency modulated (FM); also the audio is FM. In the satellite field, the audio is transported from the uplink, through the satellite, to your receive terminals as a "subcarrier." A subcarrier is an "extra" carrier, or signal, separated from the main carrier which transports the video modulation.

At the uplink transmitter, and at the output of your TVRO receiver, the video information occupies "base-band" from approximately "O MHz to 4.2 MHz." It is a mini-spectrum to itself. The frequency modulation spreads this baseband signal out, for transportation, to a wider spectrum. Now since the *video* information baseband spectrum "stops" at around 4.2 MHz, it is possible to add in some additional information (modulation), if the added information functions on a frequency *higher than* the video information. **Figure 2.0**

At the uplink, additional carriers are created. Typically, they will be on 6.2 or 6.8 MHz, although in practice, worldwide *any frequency* between roughly 5.0 MHz and 8.0 MHz can be used for this purpose. In fact, several can be generated and carried along with the video signal. Russian gorizont transmissions often carry as many as four separate "audio subcarriers," one dedicated to each of four languages that may accompany the video program. See **Figure 2.1.**

At the receiver, the "higher-than-video" sub-carrier frequencies are present through the full receiver, right down to the output of the detector. There, using special combinations of low and high pass filters, the audio subcarriers are finally separated from the video. They are sent to a second detector, one that is specially tuned to extract their audio information from the frequency on which they operate; 6.8 MHz for example.

Most receivers supply a *tunable audio subcarrier detector.* This is actually a small FM (audio) receiver that tunes from say 5.5 to 7.5 MHz and as you tune the knob, you are tuning through that frequency range "searching for" audio subcarriers. Others supply two or more "present" audio subcarrier outputs and a panel switch that allows you to select which one of the present subcarrier frequencies you will hear.

Most of the cable programming services transmit their (program) audio on the "standard" subcarrier of 6.8 MHz. Many of the network feeds use the second "standard" subcarrier" of 6.8 MHz for audio. Newer CATV multiple-audio-channel services, adding "stereo" to the satellite transmission, will use two or more audio subcarriers for that purpose, offering standard (non-stereo) audio on one subcarrier (6.8) and stereo audio on a pair of other subcarriers. (INTELSAT/GORIZONT feeds have their own "stand-ard" feed for audio subcarriers; 5.8 MHz (for western nations), or 7.4 MHz (for eastern block nations). A tunable audio subcarrier detector is recommended if you want to be sure to be equipped for all of the audio that may be "up there!"

SUBCARRIERS ON THE SATELLITES

SATELLITE	AUDIO FORMAT FOUND (MHz)
SATCOM 2R	No Video
GALAXY 2	No Video
COMSTAR 1 & 2	No Video
WESTAR 2	6.2/6.8
SATCOM 4	5.58/5.76/6.2/6.8

COMSTAR 3	5.8
WESTAR 3	6.2/6.8
TELSTAR 301	5.8/6.2/6.8
WESTAR 4	5.8/6.2/6.8
ANIK-D	5.4/5.48/5.76/5.88/5.94/6.12/6.17/6.8
ANIK-B	6.8
ANIK-3	No Video
SATCOM 1 & 2	5.94/6.2/6.8
WESTAR 5	5.58/5.76/5.8/6.2/6.8
COMSTAR 4	5.58/5.76/6.2/6.8
SATCOM 3R	5.4/5.58/5.76/5.8/5.94/6.12/6.2/6.3/6.43/6.48/6.62/6.8/7.38/7.56/7.69/7.7/7.78/7.92/8.0/8.14
GALAXY 1	5.58/5.76/5.8/6.2/6.8
SATCOM 1R	5.94/6.2/6.8
AURORA	5.8/6.8

The standard frequency plan provides up to four *stereo* channels, designated 1 through 4, in the baseband spectrum from 5.58 MHz to 7.56 MHz. The 6.8 MHz region is reserved for the television audio subcarrier. Other channel frequencies can be provided for special applications.

WHERE TO FIND THE AUDIO SUBCARRIERS

An accurate up-to-the-minute listing of all video and analog subcarrier audio programming services on the North American ("C") band satellites is published six times a year, Jan-Feb, March-April, etc., by Westsat Communications — Satellite Channel Chart—- Address: Box 434, Pleasanton, CA 94566. One year subscription rate is $19 (U.S.), $24 for international airmail. This publication gives all programming on the North American satellite channels, plus a very accurate listing of all active audio subcarrier services on each satellite. **Figure 2.2.**

COOP'S SATELLITE DIGEST carries a monthly feature column called "Transponder Watch," which carries an up-date to subcarrier and other transponder activity. CSD, P.O.Box 100858, Ft. Lauderdale, FL 33310 — $75 per year, 12 issues CSD plus 12 issues CSD/2 First Class Mail USA.

SAT GUIDE'S yellow page section lists the main video satellites, along with each satellite's subcarrier auxiliary services each month. The majority of these services are in the audio subcarrier area. The information lists the name of the service, programming category, format, transmission form, along with other information.

A basic list of audio subcarriers is given in the following list:

NON-VIDEO AUDIO SUBCARRIER SERVICES (SbC/FM)

SATCOM 3R

TR-2 Satellite Radio Network — 6.2 Mono

TR-3 Moody Broadcast Network — 5.4/7.92 Stereo

Satellite Music Network — Starstation — 5.58/5.76 Stereo

Satellite Music Network — Country — 5.94/6.12 Stereo

WFMT (FM) Chicago — Classical — 6.3/6.48 Stereo

Bonneville's Beautiful Music — 7.38/7.56 Stereo

Seeburg's Lifestyle — Music — 7.69 Mono

Satellite Music Network/Stardust — 8.05/8.14 Stereo

TR-6 Music in the Air — Country/Western — 5.4/5.94 Stereo

Music in the Air — Broadway/Hollywood— 5.58/5.76 Stereo

Music in the Air — 50's/60's—6.43 Mono

Music in the Air — Comedy — 7.69 Mono

Music in the Air — Big Band—7.785 Mono

TR-7 ESPN Information Network — Schedules— 6.2

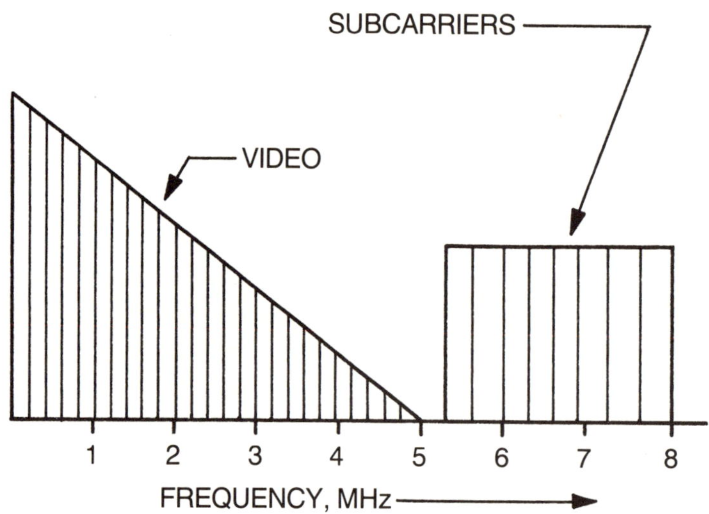

Figure 2.0 *The standard subcarrier frequency plan. 6.8 MHz for video sound with 8 more 15 kHz audio channels.*

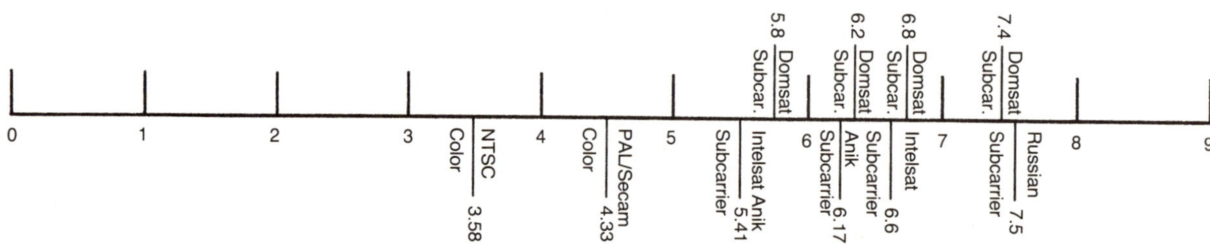

Figure 2.1 *The subcarrier spectrum.*

TR-8 Nice and Easy — Music — 5.58/5.76 Stereo
Cable Jazz Network — Jazz — 5.94/6.12 Stereo
Love Radio Network — Christian Music — 6.30/6.48 Stereo

TR-14 CNN Radio Network — Newsfeed — 6.3

SATCOM 1R

TR-20 AFRTS Radio Network — 5.94 Mono

GALAXY 1

TR-7 CNN Radio Network — Newsfeed — 6.3 Mono

WESTAR 5

TR-60 (11) MUSAK — Background Music — 5.94 Mono

ANIK-D

TR-4B (8) CKO-FM — News Radio — 6.17 Mono
CHFI/FM — Music Radio — 6.8 Mono

TR-7B (14) CKAC-AM — French Music — 5.41 Mono
CITE/FM — French Music — 6.17 Mono

TR-8B (16) CBBK/FM — CBC Radio, English — 5.4/5.58 Stereo
CBOF/FM — CBC Radio, French — 5.76/5.94 Stereo

TR-9B (18) CIRK/FM — Rock — 6.17 Stereo

TR-10B (20) CBM/AM — CBC Radio/English — 6.12 Mono

TR-12B (23) CBOF/FM — CBC Radio/French — 5.4/5.88 Mono
CBBK-FM — Radio-English — 5.76/5.94 Stereo

SATCOM 4

TR-3 Rhythm and Blues — Jazz — 5.4/6.3 Stereo
Rock-A-Robics — 40 Rock — 7.38/7.56 Stereo
In Touch — Reading Service — Mono
Georgia State Radio Network — 7.69 Mono

TR-7 Family Radio Network — 5.58/5.76 Stereo
Astro Radio Network — Religious — 6.3 Mono
Gold Mine Radio Network — Country-Western — 7.92 Mono

TR-17 Satellite Jazz Network — Jazz — 5.58/5.76 Stereo

EQUIPMENT TO RECEIVE STEREO AUDIO SUBCARRIERS

Most modern satellite receivers now come equipped with tunable subcarrier control that will cover the entire audio subcarrier spectrum, 5 to 8 MHz in the monaural code, so the tuning of any subcarrier in mono mode is a simple matter of knowing what satellites and its transponders carry the audio subcarriers as an auxiliary service. Refer to the subcarrier list.

If your TVRO receiver does not have a tunable subcarrier section built in, it is possible to modify the receiver to include this desirable feature. Many dealers and repair depots can do this for a reasonable charge.

In the May, 1982 issue of "73" MAGAZINE, Steve Gibson had a complete feature on adding an FM subcarrier decoder to your receiver. Copies can be found in most public libraries or from a ham friend who saves his copies of "73" MAGAZINE.

The second step in audio subcarrier recovery is the addition of one of the new stereo processors. Many receiver manufacturers now produce quality stereo units at reasonable cost. Refer to **Figure 2.3** for details on the popular KLM Stereo Processor. The Drake stereo unit also produces studio quality stereo from all stereo audio subcarriers. See **Figure 2.4.**

The basic hookup of these stereo units is very simple, and is covered in all of the user's manuals. Most stereo units are hooked-up using easily obtainable audio cables with RCA fittings. The entire hookup to the satellite receiver takes only 15 minutes.

These stereo units produce exceptional sound when used in conjunction with your high quality stereo amplifiers and a good set of quality speakers. Believe me, the sound is out of this world, and well worth the added cost of any of the quality stereo processors. **Figure 2.5** shows a simple stereo hookup to your present stereo system.

SATELLITE CHANNEL CHART®

THE MOST ACCURATE AND UP-TO-THE-MINUTE LISTING OF ALL VIDEO AND ANALOG SUBCARRIER AUDIO PROGRAMMING SERVICES ON THE NORTH AMERICAN C & KU BAND SATELLITES

Vol. 4, No. 3

RCA SATCOM 3R (131°W)

Polarization: ODD—Vertical, EVEN—Horizontal

TR—1	**NICKELODEON**—premium children's programming (6.8)
	ARTS & ENTERTAINMENT (Alpha Repertory Television Service)—performing and cultural arts programming (5.58 & 5.76 discrete stereo/6.8 mono)
TR—2	**PTL-THE INSPIRATIONAL NETWORK** (People That Love)—religious (5.58 & 5.76 discrete stereo/6.8 mono)
TR—3	**WGN-TV,** Chicago—Midwest's leading independent station (6.8)
TR—4	**FNN** (Financial News Network)—financial/business reporting with stock market readings (6.8)
	SPORTSTIME—midwest regional pay sports network (6.8)
TR—5	**THE MOVIE CHANNEL**(East)—24 hr/day first-run movies (5.8 & 6.8 matrix stereo)
TR—6	**WTBS,** Atlanta—Ted Turner's Superstation (6.8)
TR—7	**ESPN** (Entertainment & Sports Network/Business Times)—24 hr/day sports & business news (5.58 & 5.76 discrete stereo/6.8 mono)
TR—8	**CBN CABLE NETWORK**—religious/general family entertainment (6.8)
TR—9	**USA NETWORK**— professional sporting events, Calliope, and Ovation (6.8)
TR—10	**SHOWTIME** (West)—first run movies, entertainment specials (6.8)
TR—11	**MTV** (Music Television)—Pop/Rock Video (5.8 & 6.62 matrix stereo)
TR—12	**SHOWTIME** (East)—first-run movies, entertainment specials (6.8)
TR—13	**HBO** (Home Box Office)(West)—first-run movies, sports & entertainment specials (6.8)
TR—14	**CNN** (Cable News Network)—24 hr/day news (6.8)
TR—15	**CNN HEADLINE NEWS**—CNN newsbrief headline service (6.8)
TR—16	**ACSN—THE LEARNING CHANNEL** (Appalachian Community Service Network)—educational (6.8)
	HTN PLUS (Home Theatre Network)—first-run G and PG movies (6.8 multiplex stereo)
TR—17	**LIFETIME**—health and personal improvement (6.8)
TR—18	**REUTER'S MONITOR SERVICE**—commodity/stock market information (digital video)
	EWTN (Eternal Word TV Network)—religious (6.8)
TR—19	**C-SPAN**—live coverage from the House of Representatives (6.8)
TR—20	**HOME BOX OFFICE CINEMAX** (East)—time structured HBO (6.8)
TR—21	**THE WEATHER CHANNEL**—24 hr/day national & regional weather/environmental reporting (6.8)
TR—22	**MSN - THE INFORMATION CHANNEL** (Modern Satellite Network)—general informational entertainment (6.8)
	USA BLACKOUT NETWORK—substitute sports programming for regional blackout applications (6.8)
	HBO PROMO CHANNEL (6.8)
	OCCASIONAL TRANSMISSIONS—sporting events, news and network feeds (6.2/6.8)
TR—23	**HBO CINEMAX** (West)—time-structured HBO (6.8)
TR—24	**HBO** (East)—first-run movies, sports & entertainment specials (6.8)

Audio Subcarrier Services on SATCOM 3R

TR—2	**SATELLITE RADIO NETWORK**—contemporary/religious format (6.2)
TR—3	**MOODY BROADCASTING NETWORK**—religious (5.4/7.92 discrete stereo)
	SATELLITE MUSIC NETWORK/STARSTATION—adult contemporary format (5.58/5.76 discrete stereo)
	SATELLITE MUSIC NETWORK/COUNTRY COAST-TO-COAST—modern country format (5.94/6.12 discrete stereo)
	WFMT (FM) Chicago—fine arts/classical radio (6.3/6.48 discrete stereo)
	Bonneville's **"BEAUTIFUL MUSIC"** (7.38/7.56 discrete stereo)
	Seeburg's **"LIFESTYLE"** Music (7.695)

continued on next page

Figure 2.2 *Westsat satellite channel chart published every other month.*

Figure 2.3 *KLM Stereo Processor*

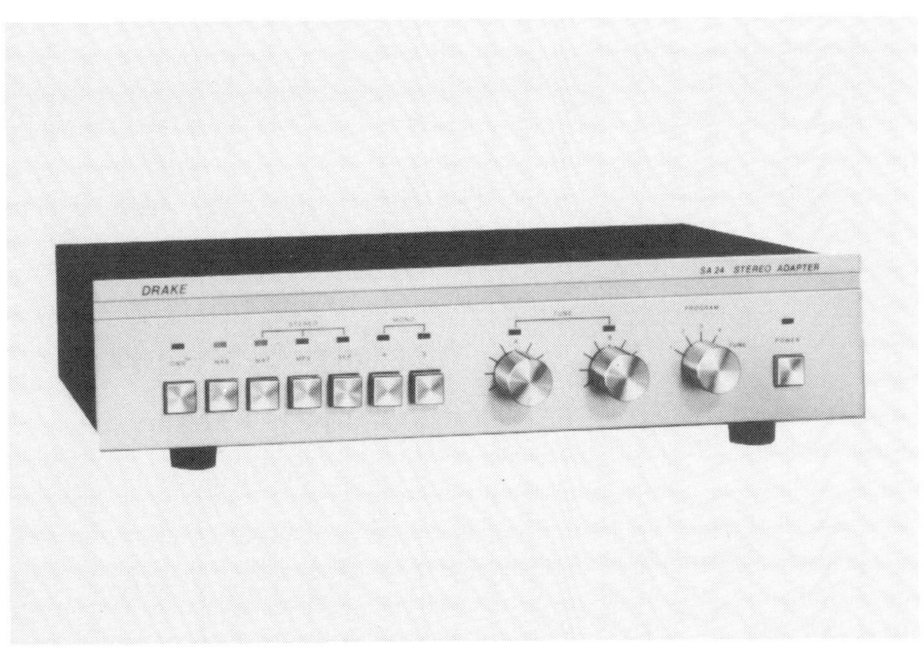

Figure 2.4 *Drake Stereo Unit*

The latest trend in TVRO receivers is the self-contained stereo processor, satellite receiver with stereo audio subcarrier all in one package. Several of the manufacturers of complete units are:

BOWMAN SR–2500
INTERSAT IQ–160 SYSTEM
VISTA–1000 RECEIVER
LUXOR SATELLITE SYSTEM
ARUNTA INVADER II SYSTEM
plus others

DRAKE SA–24 STEREO ADAPTER

This modern stereo unit is very "user" friendly and simple to hookup to almost any TVRO receiver that has the subcarrier output to make the proper connection to this unit. This must be composite video or baseband output.

If you enjoy the Nashville Network, the Movie Channel, the Disney Channel, MTV or ESPN, you can receive beautiful stereo sound to complement your great video. These networks carry stereo audio signals that can be used with your present HI-FI stereo gear. It is a real pleasure to enjoy full stereo sound that accompanies most of the most motion pictures being shown on satellite TV.

In addition to the main service carried on the subcarriers, you will find many enjoyable sound and music services, such as country and western, classical, jazz, music of the 50's and 60's, plus much more.

The Drake SA–24 processes the three methods used to broadcast stereo over satellite. The first is the Matrix Method, which uses two separate subcarriers, one subcarrier for the right channel and another for the left channel.

The second method is Multiplex Stereo, which, utilizes an FM subcarrier for left *plus* right audio and a double sideband suppressed AM subcarrier for the left minus right stereo signal. Within this system is a synchronizing signal that is used for a stereo demodulator reference.

The third system is the Discrete Method, and in this system, the right channel is transmitted on one subcarrier, the left channel is on another subcarrier, with a third subcarrier containing the audio intelligence.

The Drake SA–24 unit processes the three stereo systems, plus the monaural as well, and has the following front panel controls. See **Figure 2.6.**

DNR —	Dynamic Noise Reduction System to Reduce Background Hiss
NAR —	Narrow Filter on IF
MAT —	Selects Matrix Stereo Decoding Circuits
MPX —	Multiplex Mode Switch
SEP —	Separate Mode — Selects Discrete Stereo Mode
MONO-A —	Selects A Tuning Control
MONO-B —	Selects B Tuning Control
TUNE A —	Subcarrier Tune Channel A
TUNE B —	Subcarrier Tune Channel B
PROGRAM —	Selects One of Four Preset Channels, or Tune
POWER-SWITCH	Switches AC Power On and Off

REAR PANEL

Power Cord	
Slow Blo Fuse Holder	
Subcarrier —	In
Subcarrier —	Out
A/L —	Audio To Left Channel Amplifier
A/R —	Audio To Right Channel Amplifier

DRAKE SA–24 SPECIFICATIONS

IF FREQUENCY	10.7 MHz
IF BANDWIDTH	150 or 500 KHz
TUNING RANGE	5.5 to 8.0 MHz
INPUT IMPEDANCE	72 Ohms
SENSITIVITY	25 Microseconds 75 Microseconds Multiplex
FREQUENCY RESPONSE	15 Hz — 15 KHz
HARMONIC DISTORTION	<1% THD

OUTPUT IMPEDANCE	600 Ohms Unbalanced
OUTPUT LEVEL	250 Microvolts, Nominal

This unit performs very well, has a nice appearance, comes with a well-written user instruction book, and all in all is a nice sounding unit.

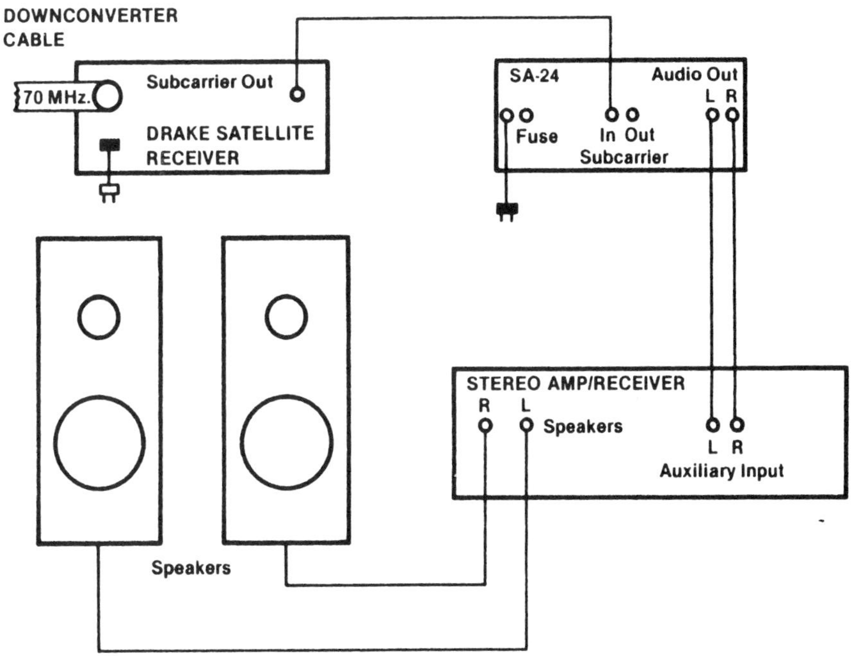

Figure 2.5 *Satellite Stereo Hookup*

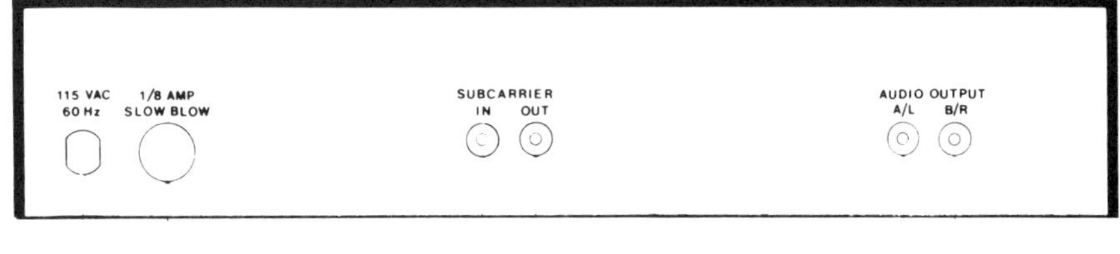

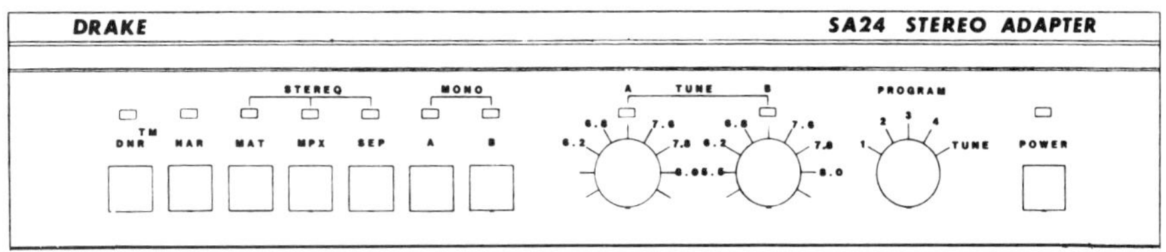

Figure 2.6 *Front controls and rear panel connections on Drake SA–24 Stereo Unit.*

MASPRO SSP–1 STEREO PROCESSOR

USS/Maspro receiver will permit you to see a brilliant satellite picture and hear the same brilliance in stereo sound.

The SSP–1's continual adjustment deviation control will also allow you to receive a vast variety of narrow to wide audio subcarrier signals that you are not able to receive on either AM or FM bands. You will open an entirely new world of audio entertainment with the SSP–1.

The SSP–1 will serve as a tuning link to your stereo system. It will allow you to use your existing amplifier and speaker system to produce the same high quality stereo sounds you are receiving from your stereo tuner, tapes, or records.

The SSP–1 utilizes three LED lights for extremely user friendly and exact tuning. Simply locate the subcarrier signal desired, (the lights indicate when the signal is most efficiently tuned). And, the SSP–1 incorporates an automatic AFC control for the lowest signal distortion possible.

To provide extremely low-noise audio reception the SSP–1 employs a noise reduction circuit that allows the total enjoyment of soft and solo performances without the annoying static and hiss commonly found with tapes, records or FM stereo reception. **Figure 2.7.**

ARUNTA SSP–318 STEREO PROCESSOR

Arunta's Stereo Processor series now in its fifth generation remains the state-of-the-art in stereo audio processing. Adaptable to any receiver the SSP–318 decodes audio transmissions in monaural, multiplex, adaptive deviation multiplex, matrix and discrete formats.

Twin frequency agile tuners cover the audio spectrum from 5.0 to 8.5MHz and bandwidth selectors with automatic deviation circuitry handle all popular ranges. The dual center tuning meters simplify audio tuning. Dynamic noise reduction, dynamic range expansion, drift-free AFC, and Canadian low pass filtering complement the advanced audio circuitry. Compact 9.5″ W x 2.5″ H x 10.25″ D size and attractive brushed aluminum and black insert face panel.

SSP–318 features are:

- Solid State IC Circuitry
- Drift-free AFC
- Fully tunable from 5.3mHz to 8.5mHz
- Loop-through composite video interface
- Multiplex stereo indicator light
- Single or Dual subcarrier demodulation
- Matrix or Discrete mixing
- Left, Right, Monaural channel outputs
- Narrow or Wide I.F. deviation selection
- Dynamic range expansion
- Canadian Filter
- Twin center tuning meters
- Automatic Volume Compensation
- DNR

See Figure 2.8.

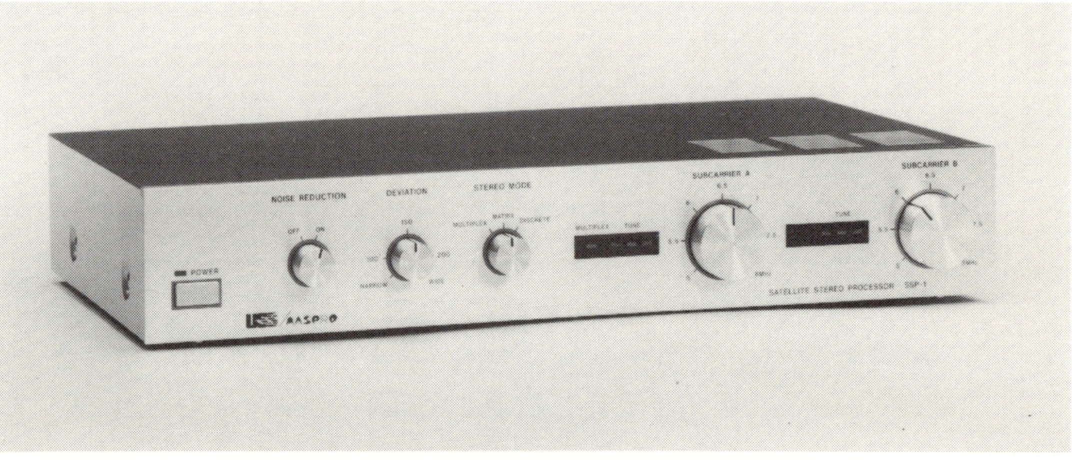

Figure 2.7 *USS/MASPRO SSP–1 Stereo Processor.*

SSP-1 STEREO PROCESSOR

Stereo Mode	DISCRETE	MATRIX	MULTIPLEX
Frequency Range		5 to 8 MHz	
Input Level		More than 50 dBm	
Deviation Control Range		75 to 300 KHz	
		(Continuously)	
Separation (at 1 KHz)	More than 50 dB	More than 40 dB	
Distortion (at 1 KHz, 75 KHz deviation)	Less than 0.05%	Less than 0.08%	Less than 0.1%
S/N (CCIR-weighted at 16 dB C/N 75 KHz deviation)	More than 50 dB		More than 45 dB
Audio Frequency Response	20 Hz to 20 KHz		20 Hz to 10 KHz
Noise Reduction		10 dB	
Power Consumption		6W at 117V AC/60 Hz	
Operating Temperature		32° to 104°F. (0° to 40°C)	
Dimensions (H×W×D)		2.76″×15.75″×9.53″ (70×400×242mm)	
Weight		Approx. 6.6 lbs. (3 kg)	

Excitement grows when you hear satellite TV stereo sound.
Courtesy, Leaming Industries

Figure 2.8 *Arunta SSP–318 Stereo Processor.*

THE FOUR BASIC STEREO FORMATS USED ON SATELLITE TRANSPONDERS

There are four basic stereo formats used on satellite transponders, plus the normal monaural sound found on most subcarriers used as the audio section for the video programming on each transponder. We will cover these four stereo formats as follows: (a) the Multiplex System — by Leaming Industries, (b) the Sum/Difference System, known as the Warner-Amex System (Matrix), (c) the Discrete Format — or Wegener System, and (d) the Times Mirror Stereo Transmission — similar to the Wegener System, except that the Wegener System uses very narrow deviation. The Times Mirror System will be treated as a basic Wegener System — Discrete.

STEREO SYSTEMS — FORMAT

SYSTEM	FORMAT	NAMES USED
Leaming Industries	Multiplex	Multiplex
Warner Amex System	Sum/Difference	Matrix SEP – A and A +
Wegener System	Discrete	Discrete
Times Mirror System	Discrete	Discrete

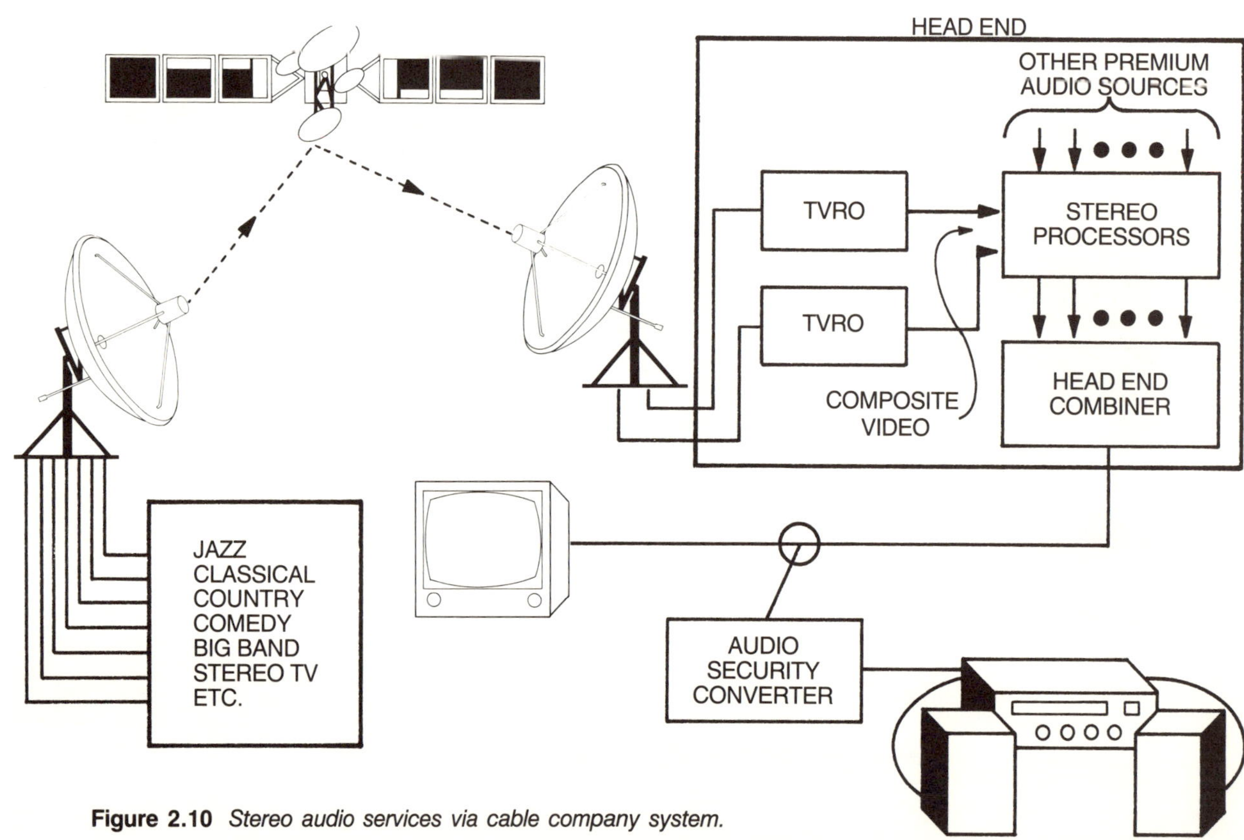

Figure 2.10 *Stereo audio services via cable company system.*

STEREO AUDIO SUBCARRIER SERVICES

SERVICE	SATELLITE	TRANSPONDER	SUBCARRIER FREQUENCY	STEREO FORMAT
Arts and Entertainment	SATCOM 3R	1	5.58/5.76	Discrete
PTL	SATCOM 3R	2	5.58/5.76	Discrete
The Movie Channel (East)	SATCOM 3R	5	5.8/6.8	Matrix
ESPN	SATCOM 3R	7	5.58/5.76	Discrete
MTV	SATCOM 3R	11	5.8/6.62	Matrix
HTN	SATCOM 3R	16	6.8	Multiplex
Moody Broadcast Network	SATCOM 3R	2	5.4/7.92	Discrete
Satellite Music Network–Star Station	SATCOM 3R	3	5.58/5.76	Discrete
Satellite Music Network–Country	SATCOM 3R	3	5.94/6.12	Discrete
WFMT/FM Chicago–Classical	SATCOM 3R	3	6.3/6.48	Discrete
Beautiful Music	SATCOM 3R	3	7.38/7.56	Discrete
Satellite Music Network–Stardust	SATCOM 3R	3	8.055/8.145	Discrete
Music in the Air–Country/Western	SATCOM 3R	6	5.4/5.94	Discrete
Music in the Air–Broadway/Hollywood	SATCOM 3R	6	5.58/5.76	Discrete
Star Ship Stereo	SATCOM 3R	6	—	Discrete
Nice & Easy Music	SATCOM 3R	8	5.58/5.76	Discrete
Cable Jazz Network	SATCOM 3R	8	5.94/6.12	Discrete
Love Radio Network	SATCOM 3R	8	6.30/6.48	Discrete
Nashville Network	GALAXY 1	2	5.58/5.76	Discrete
Disney Channel	GALAXY 1	4	5.8/6.8	Matrix
The Movie Channel	GALAXY 1	14	5.8/6.8	Matrix
Disney Channel	GALAXY 1	24	5.8/6.8	Matrix
CMTV	COMSTAR 4	18	5.58/5.76	Discrete
Disney Channel	WESTAR 5	10	5.8/6.8	Matrix
Disney Channel	WESTAR 5	12	5.8/6.8	Matrix
The Nashville Network	WESTAR5	17	5.58/5.76	Discrete
Arts and Entertainment	WESTAR 5	23	5.58/5.76	Discrete
BET	WESTAR 5	24	5.58/5.76	Discrete
CBBK-FM English Music	ANIK D	16	5.4/5.58	Discrete
CBOF-FM French Music	ANIK D	16	5.76/5.94	Discrete
CIRK-FM Rock	ANIK D	18	6.17	Multiple
CBOF-FM French	ANIK D	23	5.4/5.88	Discrete
CBBK-FM English	ANIK D	23	5.76/5.94	Discrete
BRAVO	SATCOM	2	5.8	Multiple
SPN	SATCOM 4	3	5.58/5.76	Discrete
Rhythm & Blues	SATCOM 4	3	5.4/6.3	Discrete
Rock-A-Robics	SATCOM 4	3	7.38/7.56	Discrete
Star Ship Stereo	SATCOM 4	3	—	Discrete
Family Network	SATCOM 4	7	5.58/5.76	Discrete
Satellite Jazz Network	SATCOM 4	17	5.58/5.76	Discrete

MATRIX — MULTIPLEX AND DISCRETE FORMATS

Matrix Format

The Matrix (Warner-Amex) uses a dual subcarrier system for transmission of the stereo services. Left channel plus right channel audio is transmitted on one subcarrier and the left channel minus the right channel audio is transmitted on another subcarrier. Peak deviations of ± 200 kHz are used on each subcarrier. The main carrier deviation by the subcarrier is about 2 MHz.

The two subcarriers each employ 75 microseconds of pre-emphasis and de-emphasis. The design of this system is such that remodulation on FM with 75 microsecond pre-emphasis will make the signal almost identical to a dolby-FM broadcast. See **Figure 2.11** for a typical CATV receiving set-up.

All of the stereo processors do a good job of processing the matrix stereo signals on the transponders. Most units have three controls; you would select matrix (MAT) on the popular Drake SA–24 unit.

Multiplex Format

The Multiplex Format (Leaming Industries) provides two audio channels on a single multiplexed subcarrier within the subcarrier region.

The audio baseband is of the conventional FM stereo broadcast format, with additional constraints. The left plus right (L + R) audio is sent from 50 to 15 kHz. The left minus right (L − R) is then transmitted as a double sideband suppressed carrier signal centered at 38 kHz; also transmitted is a 19 kHz pilot/signal. This pilot/signal eliminates the need to regenerate a pilot at the receive end. The deviation on the subcarrier is ± 400 kHz; this composite multiplex subcarrier is then used to modulate the main carrier levels of 3 to 4 MHz peak deviation.

Signal reception then goes through these steps: First, demodulation of the composite wideband subcarrier, then baseband composite de-emphasis and re-modulation of the FM band carrier by composite signal.

A switchable "whistle filter" is used to eliminate any audible beat notes that could occur between the 15.734 kHz sync multiples and the 19 kHz pilot/signal. This "whistle filter" has deep notches at the baseband at 15.734, 31.468 and 47.202 kHz points to prevent beats from forming.

The Multiplex System is used widely on the 5.8 and 6.8 MHz segment by Home Theater Network (HTN) SATCOM 3R, TR–16 at 6.8 MHz subcarrier; also by BRAVO Service on SATCOM 4, TR–4 at 5.8 MHz subcarrier. See **Figure 2.12,** Leaming Composite Subcarrier Stereo System.

DISCRETE FORMAT Wegener Communications

This popular format was introduced in 1981 and has found wide use in the industry. The basic system transmits a discrete left and right audio as individual low level subcarriers. These subcarrier levels, deviation and spacing have been optimized for good spectrum efficiency, consistent with full range fidelity program quality. With the Wegener System, it is possible to transmit eight monaural or four stereo pairs in the subcarrier area along with the video and the regular 6.8 MHz program audio.

A look at SATCOM 3R, Transponder 3 and Transponder 6 will reveal many of these discrete stereo subcarriers in use. Since the levels of these subcarriers are 4 to 6 dB lower than the recovered 6.8 MHz program audio, the spacing is only 180 kHz and subcarrier deviations of only ± 50 kHz are used.

The conventional TVRO subcarrier demodulator does not provide satisfactory reception of these low levels signals, consequently a stereo processor must be used to optimize the signal. See **Figure 2.0** for this Standard Frequency Plan. The Wegener System is the Series 1600 Audio Transmission System. For further information, see **Figure 2.13** Standard Discrete Frequency Plan; also **Figure 2.14** on the Wegener 1600 FM Stereo Transmission System.

In summary, in the hookup of any of the systems covered in Chapter One, an important point must be made. The type of video output from the satellite receiver must be the correct one. What is this? There is more than one type of video output? Yes, there can be — it all depends on your receiver. The correct video output or port is the one that contains *the entire baseband video in an unfiltered, unclamped condition with video de-emphasis.* Such a port is usually called and labeled *composite out.* Some receivers have two video out ports — as we said, the correct one is *composite video out.*

If there is a video out port, this will have video with clamping and video de-emphasis. Also, this port will contain additional filtering to roll off the high end of the baseband above 4.2 MHz.

If in doubt, consult the Owner's Manual or your dealer to determine which connection will give you the correct *composite video out.* When a receiver does not have a proper composite video out port, one can be added to the back of your set by any dealer in a few minutes with several small parts. You may also want to consult the receiver manufacturer if your dealer is not close at hand. See **Figure 2.15.**

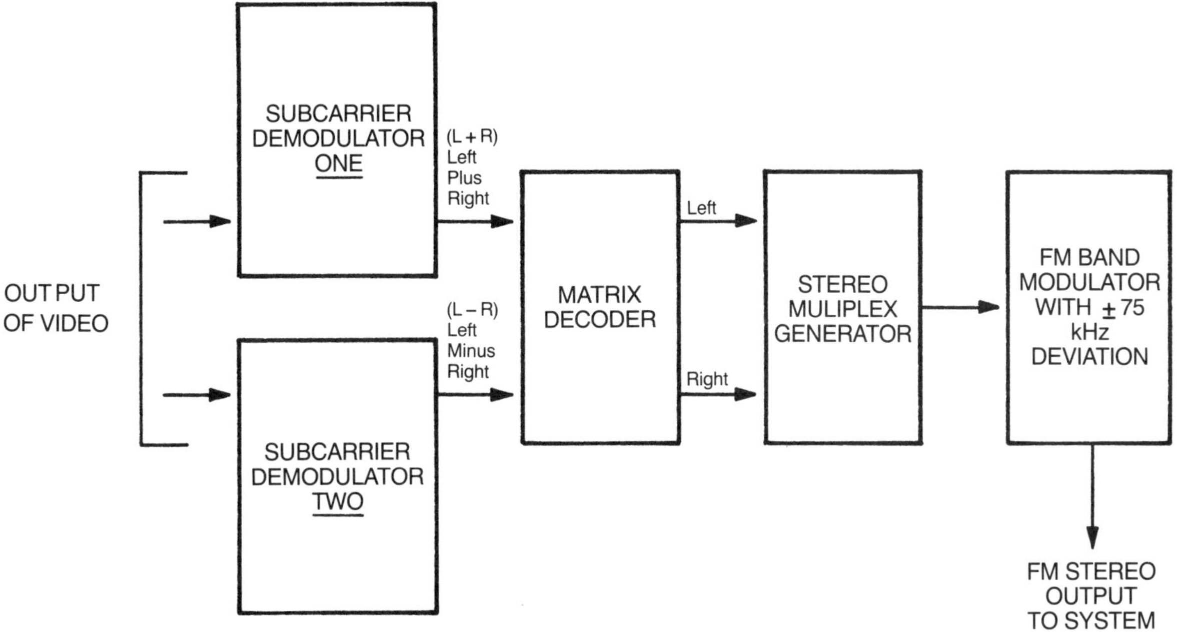

Figure 2.11 *Warner Amex Stereo System used in a CATV plant.*

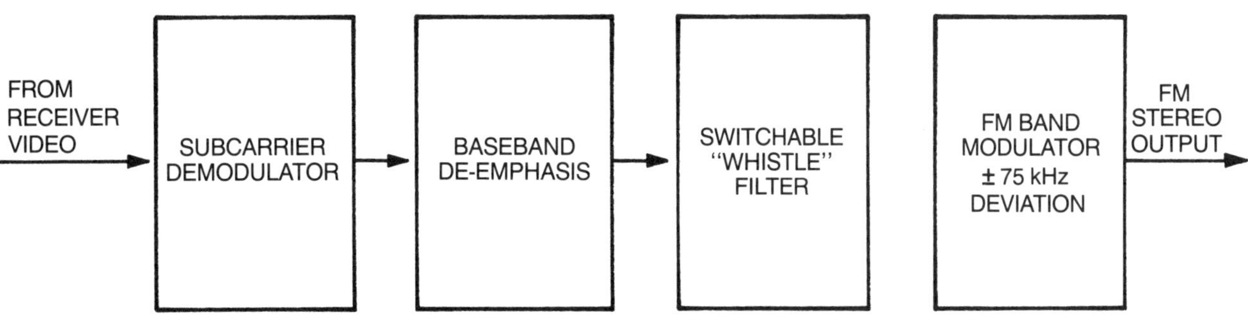

Figure 2.12 *Learning composite subcarrier stereo system.*

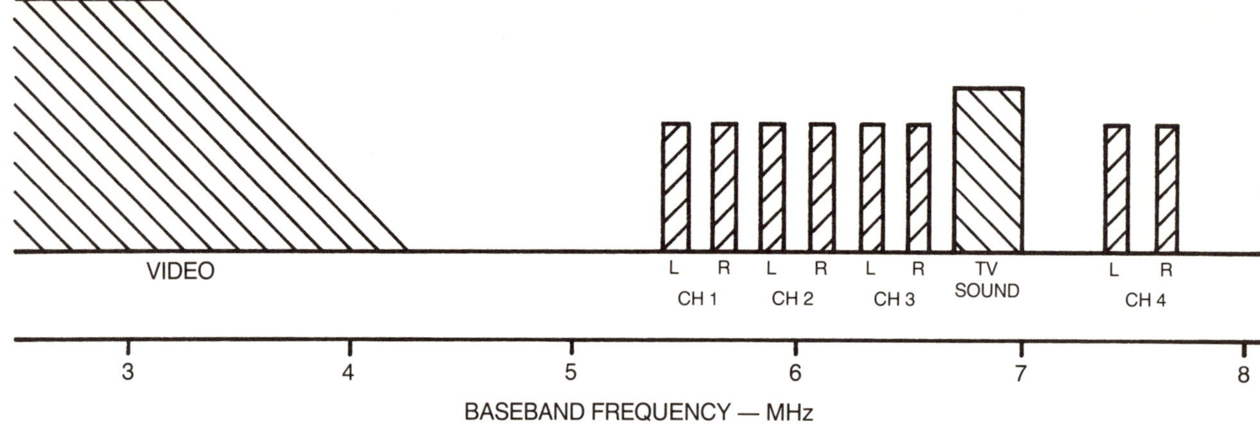

Figure 2.13 *Standard stereo discrete subcarrier plan.*

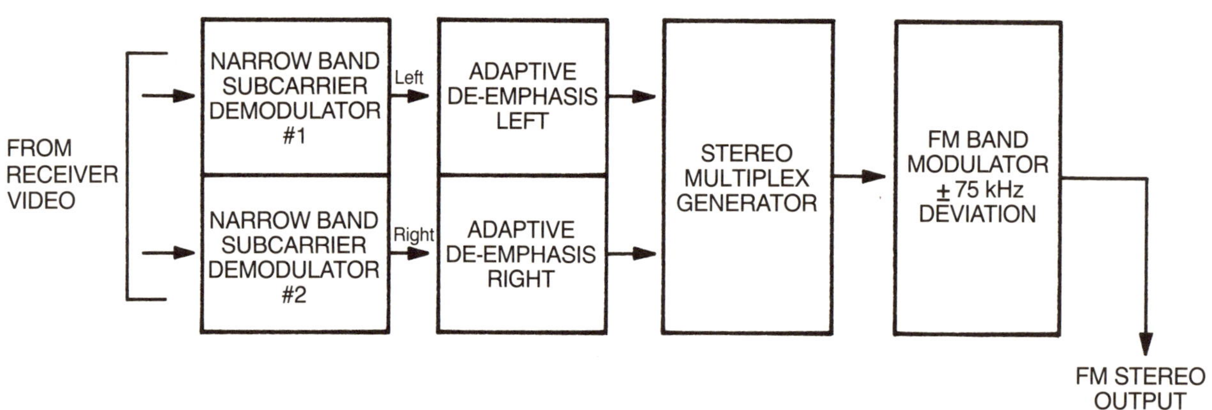

Figure 2.14 *Wegener 1600 FM stereo transmission system (discrete).*
Courtesy Wegener Communications

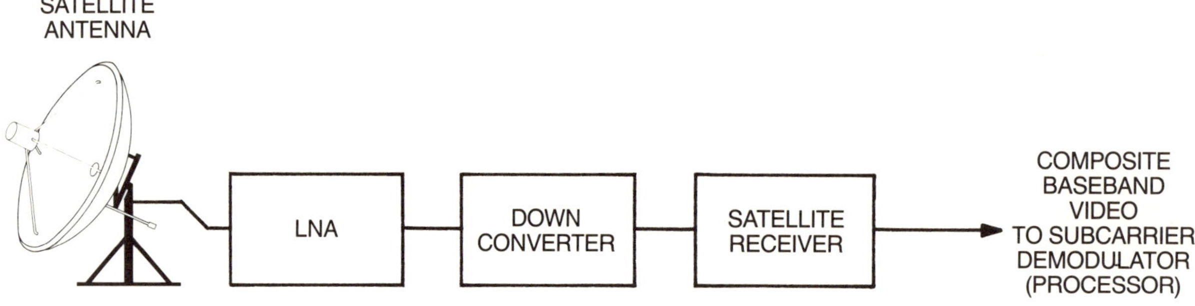

Figure 2.15 *The basic satellite receiver must provide wideband baseband signal.*

OPTIMIZING SUBCARRIER FOR SATELLITE TRANSMISSION

ABSTRACT

The use of subcarriers to deliver auxiliary services has evolved rapidly in the cable television satellite distribution network. Employing spectrum efficient transmission techniques when adding subcarriers over video can greatly increase the information throughput of the satellite channel while minimizing impact on the primary video service. This paper discusses the conceptual basis and practical aspects of the new subcarrier technology.

INTRODUCTION

The CATV industry in the last six years has created a vast communications network linked by satellite that can deliver entertainment and information to over half the homes in the United States. As a result of new technology, the satellite network can be a source of even more services to the home such as high quality audio, slow scan video, and data. In today's CATV satellite distribution system, the availability of auxiliary services can provide an additional incentive for new customers to subscribe.

STANDARD SUBCARRIERS

Canada pioneered domestic distribution of video via satellite and used a single 6.8 MHz subcarrier above video to carry three multiplexed audio channels. When satellite delivered video service was introduced to the CATV industry in 1975, the single 6.8 MHz TV sound subcarrier format was adopted by the industry as the standard. Use of additional subcarriers to provide auxiliary services soon followed. Southern Satellite Systems introduced the UPI Newstime service in 1978 using two subcarriers in addition to the standard 6.8 MHz subcarrier.

The fact that three subcarriers could successfully be transmitted along with video led to consideration of further channel expansion. The initial expansion took the form of adding multiplexed channels to the three subcarrier scheme. Multiplexing permitted channels to be added without increasing subcarrier power, thereby improving the efficiency of subcarrier power utilization. Compandors were used in the multiplexed channels to reduce the subcarrier bandwidth. This evolution was a major step toward the concept of spectrum efficient subcarriers.

Although these techniques increased spectrum efficiency for narrow bandwidth channels, conventional multiplexing could not provide many high quality 15 kHz audio channels. This led to the development of narrow bandwidth, low deviation subcarriers for transmission of high quality audio.

SPECTRUM EFFICIENT SUBCARRIERS

There is not a precise definition of a spectrum efficient subcarrier but there are some guidelines that may be applied. The baseband spectrum that may be used for subcarrier transmission is limited, approximately bounded by 5.4 MHz on the low end and 8.0 MHz on the high end. The subcarrier power that may be delivered into this spectrum is limited by available satellite power, earth station size, and overall RF bandwidth of the satellite channel. With this in mind, a spectrum efficient subcarrier may then be defined as one which:

a) Uses a modulation technique which is efficient for the type of information transmitted.
b) Uses signal processing techniques to improve modulation efficiency where possible.
c) Optimizes power and bandwidth of the subcarrier within the constraints of the satellite transmission system.

The following spectrum photographs illustrate the application of spectrum efficient subcarriers. All photographs are of baseband spectrums as received off the satellite and are centered on the 6.8 MHz TV sound subcarrier. In all cases, the 6.8 MHz subcarrier deviation is 2.0 MHz peaks.

Figure A illustrates the current baseband arrangement of Southern Satellite System's transponder 6 on Satcom I. From left to right, the first two subcarriers are a 15 kHz stereo channel, located at 5.58 and 5.76 MHz, using the WCI Series 1600 Audio Transmission System. The next subcarrier is a multiplexed subcarrier using the WCI 1585A Audio-Plexor. It provides an 8 kHz slow-scan video channel and a 5 kHz audio channel for the North American Newstime service. The 7.4 MHz subcarrier is also multiplexed using the WCI Series 1100 Multiplex System. It provides a 10 kHz color slow-scan video channel for "The Womens Channel" and five 5 kHz audio channels.

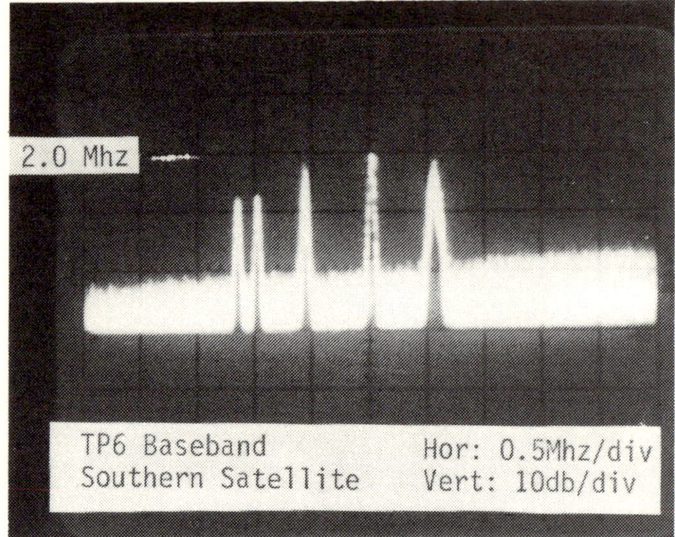

Figure A

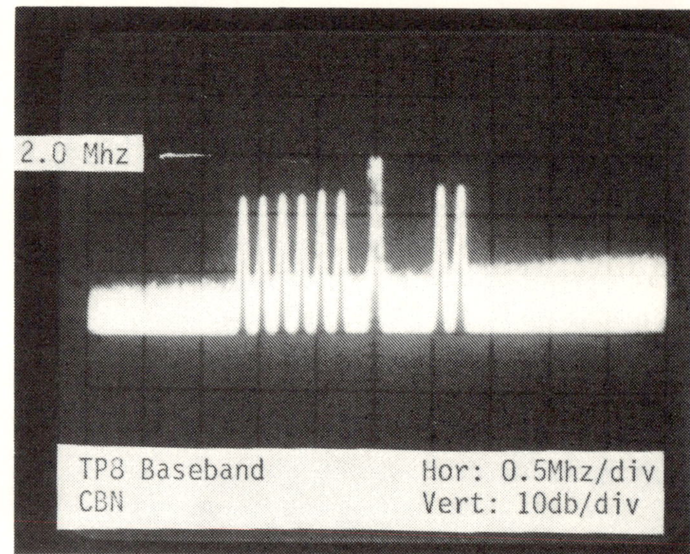

Figure C

Figure B shows the subcarrier baseband configuration of United Video's Transponder 3. At 5.8 MHz on the left is a multiplexed stereo subcarrier for WFMT. Next are two narrow-bandwidth subcarriers at 6.30 and 6.48 MHz, also providing a stereo channel for WFMT. At the extreme right is a conventional 7.6 MHz monaural subcarrier for the Seeburg Lifestyle music service.

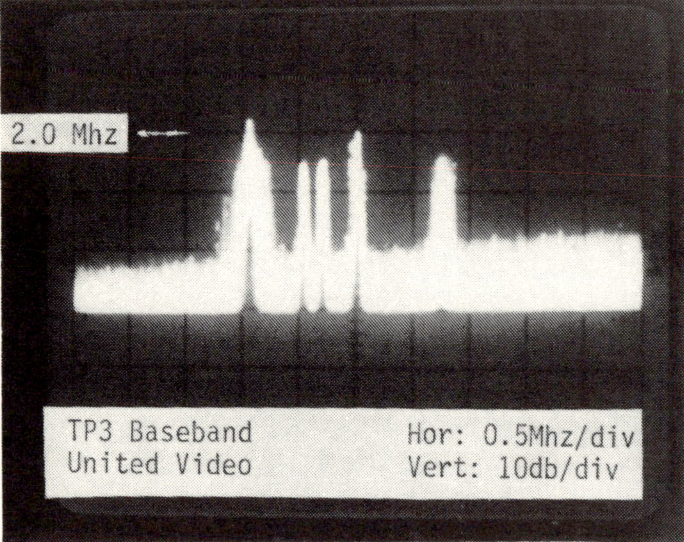

Figure B

Figure C is the current baseband of CBN's Transponder 8. It contains eight narrow bandwidth subcarriers providing three 15 kHz stereo channels and two 15 kHz monaural channels .

The narrow bandwidth subcarriers provided by the 1600 Audio Transmission System represent a major advance in high efficiency subcarrier technology. These subcarriers have been optimized for transmission of high quality 15 kHz audio channels. Using one-fourth the subcarrier power and one-half of the bandwidth of a standard subcarrier, they deliver approximately a 5 dB better signal-to-noise ratio than a standard subcarrier. This dramatic improvement in efficiency is made possible by the development of a specialized compandor which functions as an adaptive pre-emphasis system to provide 18–20 dB of signal-to-noise improvement. Because of the signal-to-noise advantage, subcarrier power and bandwidth can be reduced. The end result is a system that can deliver four very high quality stereo channels in addition to TV video and sound.

The use of a system that transmits eight additional subcarriers raises many obvious questions. On multiple subcarrier systems, especially those using equally spaced subcarriers, there were early concerns about intermodulation. Third order intermodulation products are produced on the same frequencies as the subcarriers. Second order products are produced in the video baseband with the potential of video interference. However, due to the excellent linearity of the satellite video channel, intermodulation products are very low. Third order products are typically less than 50 dB below the subcarriers and the second order products are 65–70 dB below video. These levels are well below the point at which the intermodulation products degrade the audio or video.

Another key question concerns the RF bandwidth required for transmission of nine subcarriers. An increase in bandwidth could have considerable impact on the performance of existing video receivers. To further complicate matters, it is known that the present method of predicting the RF bandwidth of a video signal with subcarriers is imperfect. Lab and satellite testing must be used for all multiple subcarrier configurations to be sure that the resultant RF spectrum will not degrade the network. **Figures D–G** illustrate the effect on RF spectrum of nine subcarriers compared to one subcarrier. **Figures D** and **E** show the spectrum produced by a color bar test signal with one and nine subcarriers. The video deviation is 10.75 MHz for the single subcarrier case and 9.75 MHz for the nine subcarrier case.

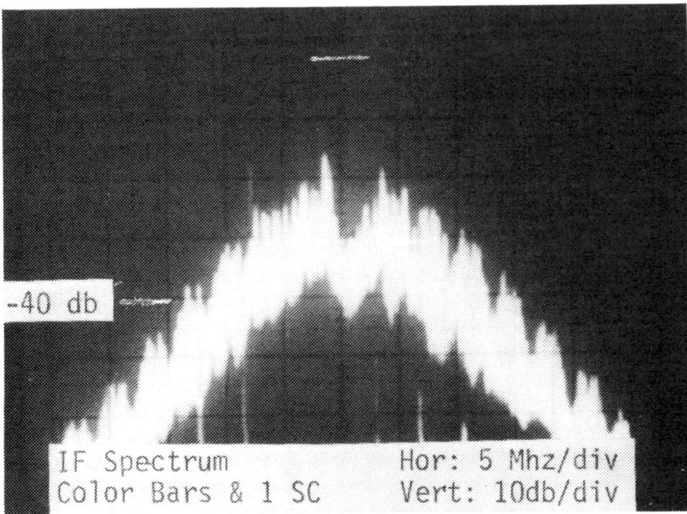

Figure D

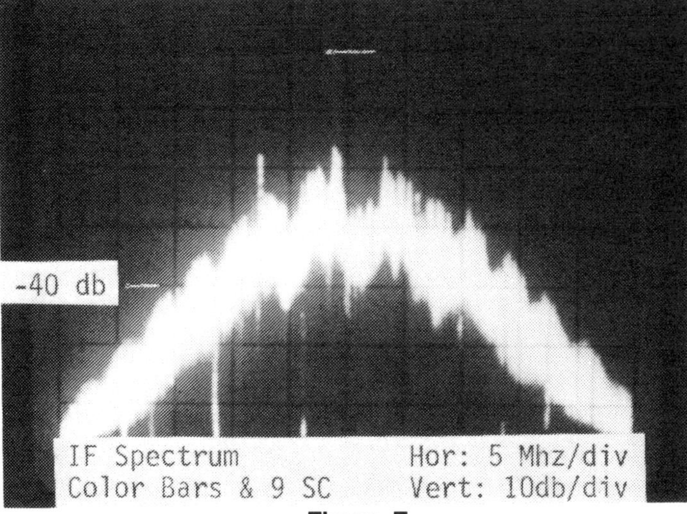

Figure E

The above photos show very little difference in the −40 dB bandwidth of the spectrum. **Figures F** and **G** illustrate the one and nine subcarrier spectrum with a typical video program.

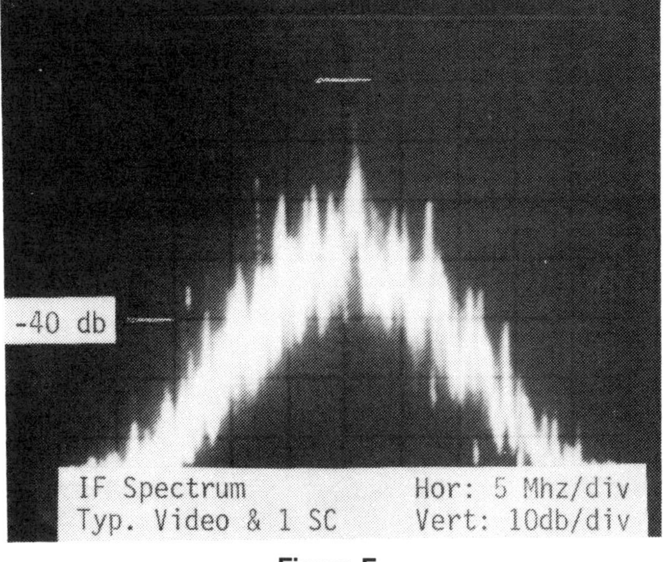

Figure F

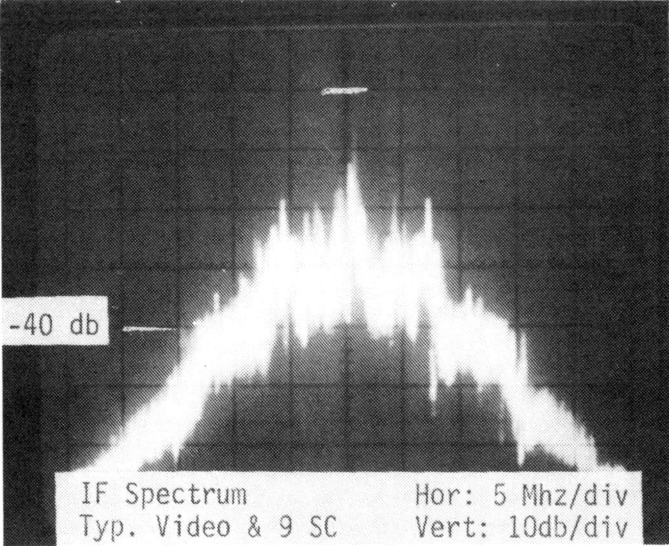

Figure G

TESTING

The spectrum characteristics shown above indicate that a nine subcarrier system does not produce a spectrum much different from a one subcarrier system. The 1600 Audio Transmission System was exhaustively tested to prove that a nine subcarrier system is compatible with the existing CATV network. Tests were conducted on six different transponders used by networks totalling nearly 30 million subscribers. In all cases, no degradation of network service was reported.

NOTE:

This abstract by Elias Livaditis, Bob Placek and Heinz Wegener. Courtesy Wegener Communications

"COOP" TAKES A FURTHER LOOK AT SATELLITE SUBCARRIERS (SbC/FM)

180 kHz AT 4 GHz

Most of the popular, often utilized video transponders transmit their audio to receive locations by 'marrying' the audio to the video signal as a subcarrier. Let's see how this works.

Our standard satellite television format sends signals into space in the 6 GHz region; the uplink. Inside the satellite these incoming signals are 'mixed' to a new set of frequencies, in the 4 GHz region, amplified, and re-transmitted back to earth through a downlink transmit antenna.

The original video signal is a baseband signal, occupying a frequency range between 0 MHz and perhaps 4 MHz. This is the sort of type of signal which you could view by connecting it to a video 'monitor.' A monitor differs from a video receiver in that a receiver tunes in the initial frequency as a *radio frequency* (RF) *carrier,* while a monitor tunes in the *information* being carried by the radio frequency carrier. It can do this at two points in the chain from uplink to you; *before* the original video signal is 'married to' a radio frequency carrier, or, at the opposite end of the chain, *after* the radio frequency carrier has been received at the downlink, and the original video information extracted from the radio frequency carrier.

A standard television receiver combines the features of a radio frequency carrier receiver, *and,* a monitor. The radio frequency portion takes the signal out of the air and amplifies the signal, converting the radio frequency carrier to a lower frequency so that the modulation or video information can then be detected or extracted from the radio frequency carrier. You can convert, or change the frequency, just about as often as you wish, and the modulation or original baseband information will stay right with the radio frequency carrier.

In the satellite television format, the transmission mode is FM or frequency modulation. This means that the radio frequency carrier is married to the video information by varying the frequency of the radio frequency carrier. In terrestrial television formats, the video is normally transmitted in an AM or amplitude modulation format. That means the amplitude or strength of portions of the signal (a sideband) are varied by the content of the modulation information (i.e. video).

A frequency modulated signal uses up more 'spectrum' or 'space' to transmit the same picture information than an amplitude modulated signal. However, for high quality transmission systems, FM has many advantages which offsets the wasteful use of spectrum.

Our satellite video and our satellite audio are all transmitted by FM. The video signal occupies approximately 50% of the width of a normal 36/40 MHz wide transponder. This leaves the remaining 50% or so of the transponder width to stick in one or more audio channels associated (or not associated) with the video being transmitted.

To more fully appreciate what is being done with the audio, it is imperative that we study the make-up of a baseband format. This would be the same format at either the uplink, before the baseband (raw video and raw audio) is married to the radio frequency carrier, or, at the downlink, after the receiver has received the 4 GHz downlink signal, converted that signal to a lower IF (such as 70 MHz), and finally demodulated or extracted the video and audio signal(s) from the radio frequency carrier.

There is a handy rule of thumb which can be used to study the way a baseband signal ends up modulating a radio-frequency carrier in the services of interest to us. Satellite engineers consider 36 MHz of a 40 MHz wide transponder useful. They look upon the remaining 4 MHz as a pair of two MHz wide 'guard bands,' at opposite ends of the 40 MHz wide transponder 'channel.' Guard bands are avoided to insure that there is not interference between transponders which are adjacent, in frequency or the spectrum, to one another. Now remember that with FM, it is the frequency which the baseband signal is modulating. That means, for ease of understanding *if not literally,* that it is the frequency of the radio frequency carrier which is shifted or moved about by the presence of the video (and audio) information. If our baseband video information occupies 4 MHz (0 to 4 MHz is the same as being 4 MHz 'wide'), and we have 36 MHz to modulate, in theory we could force the 4 MHz wide video to deviate or 'swing' the radio frequency carrier over a 36 MHz wide region. The ratio between 36 and 4 is '9'. However, we have other considerations to be concerned with, including leaving enough room to send along an audio carrier (or more than one audio carrier). Space in that 36 MHz wide channel must be 'saved' for the audio.

To save spectrum, it was determined that approximately 20 to 22 MHz of the 36 MHz wide channel would be adequate for video; and the balance would be reserved for the transmission of audio and other data. It was also determined that the most economical use of the original baseband spectrum would be in a region from 0 MHz to approximately 8 MHz. All of this worked out so that the center 20/22 MHz of a 36 MHz wide channel would be reserved for the video, while the outside edges of the transponder, another 14 to 16 MHz of transponder space, would be left for audio. It makes a little more sense if we diagram it, and if we do *not take* the previous explanation too *literally!*

The modulation process is akin to a multiplication process. We have a baseband spectrum from 0 to 8 MHz, and we have a final transponder width of 36 MHz. We have to make the 8 MHz baseband spectrum "fill" the 36 MHz wide spectrum so we end up increasing the width of each part of the original 8 MHz wide spectrum by approximately 4.5 times. And to keep the video and audio separate, at both ends, we arbitrarily assign a segment of the 0 to 8 MHz wide baseband spectrum to audio. We already know that the video is approximately 4 MHz wide, so that means that the audio will end up being above the 4 MHz width-point of the video.

Just as a full transponder requires a guard band between transponders, so too does a baseband 'transition' between the video information and the audio information. In this case our 'guard band' extends from approximately *4 MHz to approximately 5.5 MHz.* That is designed to insure that the video information does not 'crawl up' into the audio portion of the signal, and that the audio does not crawl "down" into the video. Of the two 'crawls' it is the video going up which presents the biggest engineering challenge since the video signal is 'wideband' by nature, while the audio signal is (relatively speaking) quite narrow band.

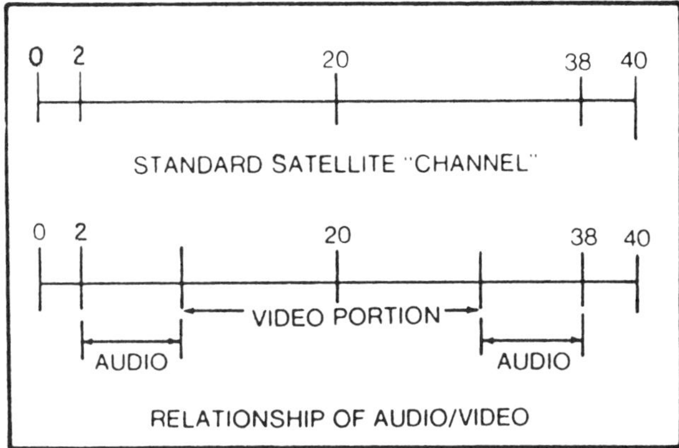

RELATIONSHIP OF AUDIO/VIDEO

The audio signal has to have a carrier of its own to ride along on, or it would wander about. Designers take the baseband frequency spectrum from 0 to 8 MHz and assign specific audio carrier frequencies to audio carriers. Because these audio carriers are secondary to the main video carrier, they are called 'sub' carriers. Common subcarrier frequencies are 5.8, 6.2, 6.8, and 7.4 MHz. You will note that these audio subcarrier frequencies are spaced apart by at least .4 MHz and can be as much as .6 MHz apart. Again, there has to be adequate separation or space *between the audio subcarriers* to keep them from interfering with one another, and to make it possible for the receiver at the downlink to separately and individually tune in the different audio 'sub' carriers.

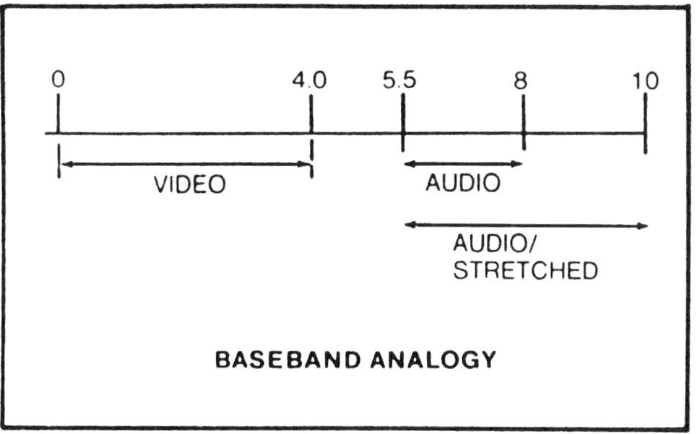

BASEBAND ANALOGY

One of the happy characteristics of an FM transmission system is that it is possible to send a subcarrier along with the video carrier for a very small additional 'price.' The entire system is designed to carry the primary or main carrier. Once everything is in place, from the uplink transmitter to the satellite, to the downlink receiver system, the audio subcarriers can be added for very little additional money. This is the primary reason why you see firms that use satellite to sell video services, such as United Video's carriage of WGN in Chicago, adding more and more audio subcarrier services. They have the full system in place, and they are paying for it with the transmission of video. *Now* they can add audio subcarrier services for a tiny fraction of the original cost, serving perhaps not very many users per audio subcarrier in the process. But because the BIG price has already been paid for the transmission of the video, the audio becomes a very economical 'add-on' service.

There is a price for adding audio of course, even if it appears to be small. There is additional equipment, and, *there is a penalty paid* by the primary or main video carrier. How does that work?

Every satellite transponder has a full rated transmitting power. For example, on F3R, WGN on transponder 3 has a maximum output power of 8.5 watts. That 8.5 watts is available to relay from space back to earth the WGN video carrier, and, the WGN audio carrier (6.8 MHz). Now, what happens if United Video adds additional subcarriers, at say 5.8 and 6.2 MHz? *Each* of these additional carriers will consume *some* of the available 8.5 watts of power. It turns out that each subcarrier uses between 0.5 and 0.75 dB of transmission power available. Or to put it another way, if *your* actual footprint from WGN was 34 dBw, with *one* audio subcarrier present (the program audio channel), when they added two additional (and non video program related) audio subcarriers, your footprint signal level would *decrease* by 2 times .5/.75 or from 1 to 1.5 dB. That is not an insignificant reduction in power since it comes close to being the difference in antenna gain between a 10 foot and a 13 foot antenna!

And when you combine a weaker-than-average footprint (i.e. WTBS) and add, as Southern Satellite System is now *planning* to do, several subcarriers to handle five or more separate audio services, the reduction in available footprint power, to your antenna, can become quite significant indeed.

So while it is small, as measured in terms of dB, there is a price paid when additional (audio) subcarriers are added to the basic video plus one subcarrier format of most video program services. If you have a TVRO receiver which affords you the luxury of tuning for audio (or data) subcarriers, you can build your own relationship 'chart' between those transponders that are strong, and those that appear somewhat weaker, as that observation relates to the presence of *more than* a single (audio) subcarrier.

STAYING LOCKED UP

One of the big advantages to the subcarrier system is that the audio carrier is locked to, or referenced to the presence of a video carrier. If your receiver is properly tuned in for the video carrier, the audio will come along on the receiver baseband at the proper 5.8, 6.2, 6.8 (etc.) spot 'on the dial.' Your receiver has either an automatic frequency control (that keeps it locked on the video carrier), or a manual control which you keep 'peaked.' *The 'video lock' insures an audio lock as well.*

A standard receiver must keep the video portion of the signal within a 30 MHz wide, or so, 'passband' at the incoming 4 GHz frequency. As a ratio, that's keeping 'one part in 133' for stability and with modern down conversion oscillators (the oscillator is where the drift occurs), that is not a tough assignment. But, not all of the audio carriers of interest on satellite these days are transmitted in that format. And that is where the problems really begin!

Take, for example, that transmission of a TV program (or other) audio channel sent separate from the video, on a transponder removed from the video. This is a particularly popular system with many of the Middle Eastern countries utilizing Intelsat for relay of internal video programming. It has also become popular, recently, with several of the new users of Intelsat from South America (Colombia, Peru). Take Saudi Arabia, as an example (see diagram). The video is transmitted as a full transponder (TR1) from 21.5 degrees west. The antenna system is an eastern hemispheric, which means the service is not seen west of the bird.

Saudi Video is centered on 3725 MHz (3.725 GHz) and the audio is five transponders higher, near the very top end of transponder 5, roughly 3900 MHz. How, with conventional home system equipment can you recover that audio?

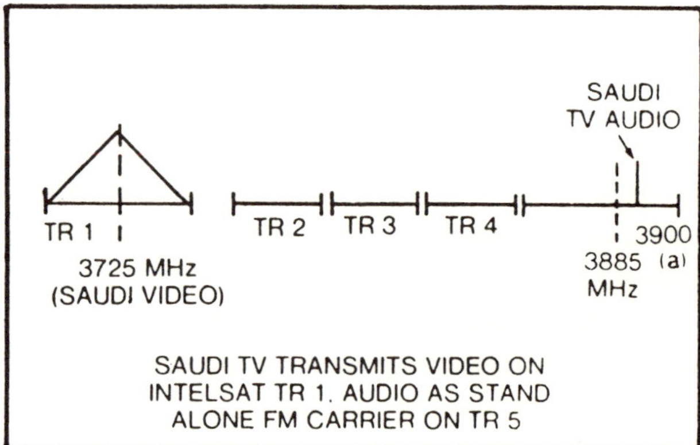

SAUDI TV TRANSMITS VIDEO ON INTELSAT TR 1. AUDIO AS STAND ALONE FM CARRIER ON TR 5

Yes, if you want video and audio at the same time (simultaneously), you will need a *pair* of receivers. One for the video, and one for the audio. Then what?" "Simply put the audio receiver on transponder 5, and tune in the audio with a tunable audio system. "No, that will not work.

Remember that a baseband tuning audio system functions because the audio is sent along with the video, as a subcarrier to the video. In the Saudi case (and others) there is no video present. Without video, there are no subcarriers.

So if Saudi is transmitting audio *without* a sub-carrier, how are they getting the audio on transponder five? By using a technique known as single carrier FM. It turns out that the Saudi's wanted their video to be as strong, and as close to perfection as is possible, with the satellite link. They had two transponders leased on Intelsat, and decided that if they took the audio *away from* the video transponder they could gain back both some extra 'power,' plus the use of a full 36 MHz for video-only. Transponder 5 *(reads 9–10 on a US receiver dial)* has not only their television program audio, but also a large quantity of Saudi telephone and data traffic transmitting in an SCPC format.

That still does not tell us how to get it back.

Demodulating the audio FM signal is a secondary problem to holding it in one place long enough to be able to tune it in and demodulate it! To explain; the automatic frequency control (AFC) in a standard TVRO receiver has to keep the receiver local oscillator within 2 to 4 MHz. The video carrier is the reference the AFC locks on, typically, and *as long as* there is a video carrier present, the AFC has *something to reference to.* Lacking that, the local oscillator in the receiver (it provides the signal to the mixer, which in turn down converts the 4 GHz signal to ultimately 70 MHz [IF]) has nothing to reference with. In the Intelsat installations, a 'pilot' or control carrier is transmitted *near* the middle of the transponder (or half transponder) as a reference signal. The receiver has a special detector that locks onto that reference carrier, and holds it within stability requirements so that the balance of the narrow audio carriers end up where they should be in the IF system.

This is a very expensive requiring phase-locked synthesized oscillators for each of the in-use audio channels. Drift, then, or keeping the 4 GHz signal inside the 'passband' of the demodulator, is the number one problem involved. We'll come back to the Saudi system shortly.

A number of countries, such as Niger, Oman, Zaire, Morocco and Algeria use a system for video transmission called 'enhanced half transponder.' See *illustration.* Each of these users places the video in the lower half of the transponder, and the audio is sent in the upper half of the transponder. The audio may also be joined in the upper half by other 'domestic' audio channels, or communication channels.

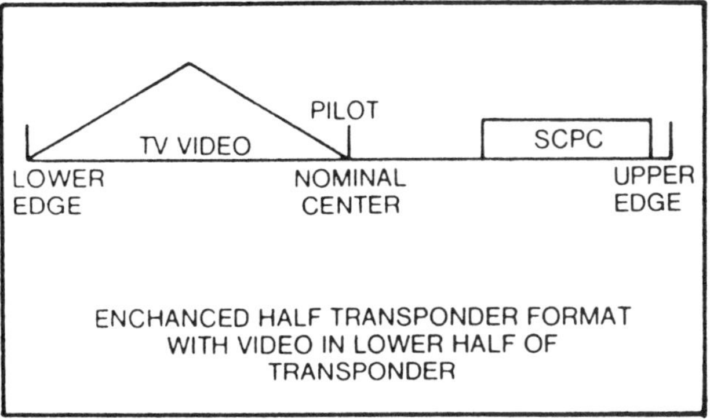

ENCHANCED HALF TRANSPONDER FORMAT WITH VIDEO IN LOWER HALF OF TRANSPONDER

Now it is possible, in either a single or double conversion system, for the receiver LO to be so placed that it can be 'below' or 'above' the signal being converted. For what follows, the receiver LO in a single conversion receiver must be *below* the signal being converted (called 'right way up' in the trade). For a double conversion system, the LOs (there are two) must *both* be below, *or,* both be above the signal being converted. 'Erect' is the term given to right-side up conversion and 'inverted' is the term applied to reversed LOs.

Now if you are tuning in any of the 'lower half/half-transponder format' services which transmit the video at the bottom of the transponder and the audio at the top, you tune your half transponder format receiver so that the half transponder wide video signal is centered on 70 MHz. That's the normal tuning procedure. Now, if the video is centered at 70 MHz, and the video is really half way down the lower half of the transponder, the upper half of the transponder, which contains the audio carrier, will now be up there between 88 and 100 MHz. And that suggests a method of copying the audio since the transmission format is FM, and standard FM broadcast receivers tune from just below 88 MHz to 108 MHz.

To tap into this signal range, you have to be mindful that in a half transponder format receiver (such as the AVCOM or ICM) the IF bandwidth is on the order of 15 to 20 MHz; or about half the normal bandwidth. If the IF system if properly designed, that will allow signals within the IF strip from approximately 62.5 (60) MHz up to 77.5 (80 MHz) to pass through to the video detector. Those 'way up there' *at 88 MHz and above* will be 'clipped' or severely attenuated by the IF filter. Yet you have to get your FM tuner/receiver connected into the 'loop' if you are going to tune in these audio FM carriers.

One technique employed with a measure of success is to look carefully at the 70 MHz IF amplifier 'string.' The filter may fall at the input, in the middle, or at the output. Each design has merits. Most often, however, the filter will end up *in the middle* of the 70 MHz string. This suggests that if you go into the IF board with a tap-off point inside of the IF, *but ahead of the bandpass filter,* you can derive sufficient signal to drive out through a short piece of coaxial cable to the outboard 88–108 MHz tuning receiver/tuner. Some have found that there is sufficient signal at the input to the IF, ahead of the IF amplifier/filter string and by merely adding a back matched or hybrid two-way signal splitter here (common 75 ohm CATV device), you can drive the IF out of one port and the outboard FM tuner/receiver through the opposite port. The danger here is that you may have a marginal amount of signal coming from the down converter to the IF, and a 3.5 to 4 dB reduction (caused by adding the signal splitter *ahead* of the IF) may reduce the quality of the video picture. It is worth a quick try, however, with two piece receivers *as long as you check* to be sure the IF line is *not* carrying a tuning or operating voltage for the remote down converter. You don't want to try adding a signal splitter in the line where there is tuning or operating voltage present!

Now, having made the connection, how does it work?

The half transponder signal is centered on the 70 MHz IF by the fine tuning control on the receiver. The receiver AFC references to that signal, and holds it in place to perhaps a couple of MHz if the AFC is good. *But there will be drift.* The 4 GHz range oscillator is simply not stable enough under varying heat and operating conditions to be absolutely stable.

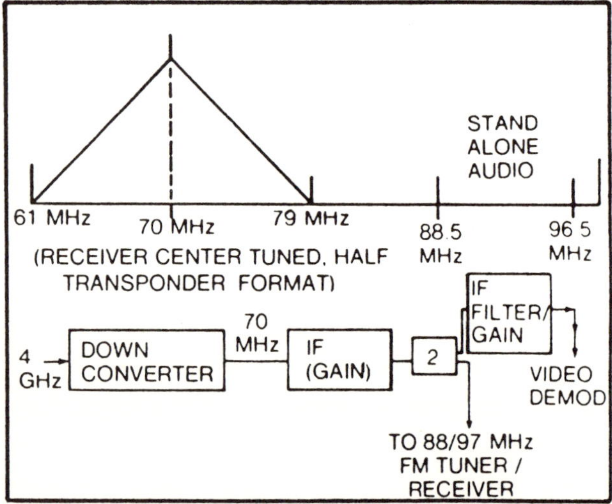

This instability will not be noticed on the video since the 'wideband' FM TV signal is fairly tolerant of such changes. But, when you have spotted Algerian audio at say 95 MHz, and have it tuned in, when the LO moves 2 MHz, the Algerian audio will also move (on your FM tuner/receiver) by 2 MHz. Such a system would possibly not be a suitable method of extracting the strange stand-alone FM audio carriers for a commercial installation, but it does give other terminal users at least access to these otherwise unavailable services. There are other, hypothetical, ways to accomplish the same thing but they involve engineering hardware which does not, apparently, exist at the moment. We'll return to this, also, shortly.

Now that we understand that there are some services which separate the video and audio, and the audio arrives at the downlink site via a method other than subcarrier, what about the unusual technique employed by Saudi Arabia? Well, it turns out that where countries have leases on two or more transponders (or parts of two or more transponders), they may well opt to do this. The recent Colombian video addition to Intelsat apparently is transmitting their audio on a transponder 14 away from their video!

One solution is to use two separate receivers, with a splitter from the LNA output feeding the main receiver. One receiver, in the case of Saudi, would tune in the video on US equivalent TR1 while the second would tune in the audio on the US equivalent of TR9. If the TR9's receiver is one of the half transponder receivers, by setting the receiver fine tuning so that the *lower* half of the transponder is centered at 70 MHz, then you will be able to use the previously described FM tuner/receiver play to tune in the audio within the FM band. (In the case of Saudi, you will also find some radio program channels transmitted in the *same* 'upper half' *of the transponder,* but they will obviously not match the video, and for video-audio reception can be bypassed).

As noted, the receivers depend upon the presence of a video carrier to lock the AFC to *something.* Without any video on the lower half of the Intelsat TR5, the receiver is going to try to *find* something to lock to. There *are* carriers on the low end of that transponder (although not video), and the AFC *may* try to latch to one or more of them. When that happens, your carefully tuned in Saudi audio is going to move away from the spot on your FM dial where you had tuned it in!

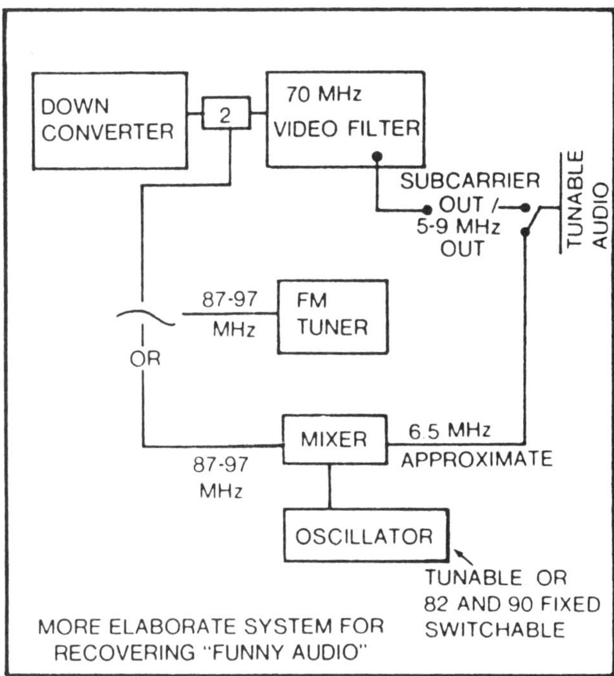

MORE ELABORATE SYSTEM FOR RECOVERING "FUNNY AUDIO"

THE NEXT STEP

There are several techniques being investigated to 'clean up' the quick and perhaps dirty approach of tapping into the 70 MHz IF with an FM tuner/receiver. For those who may have some talents in such areas, these proposals are offered:

1) *Convert the audio* directly to a 10.7 MHz IF. Virtually all of the FM tuners/receivers utilize a 10.7 MHz IF, and one way to take advantage of their selectivity and still get the package stuffed into a modified TVRO receiver would be to build a tunable VCO which tuned either 76–86 MHz (or 98–108 MHz; *this* would keep the VCO *out* of the range normally covered by the TVRO reciever IF), and 'mix' the audio carriers directly into a 10.7 MHz IF. This leaves the control that tunes the VCO as a 'tuning control' and allows the user to tune, directly at the 70 MHz IF, the strange stand-alone FM carriers.
2) *Convert the audio* directly to the 6.8 MHz position on a fixed tuned TVRO receiver. The approach here is the same as with the 10.7 MHz IF, except that the VCO will now tune 80–90 (or *preferably* 94–104) with the mix product being at 6.8 MHz. This means the audio system already built into the TVRO receiver will become a part of the system, and user will simply 'insert' the additional VCO/mixer into the line when he wishes to employ 'direct SCPC tuning' for stand alone, wideband FM carriers.
3) *Convert* the 87–97 MHz region down to 3 to 13 MHz, and modify the tuning range of existing TVRO receiver audio demodulators to tune this range.
4) *Leave* the existing 5 to 9 MHz (nominal) TVRO sub-carrier tuning range alone, but install a fixed oscillator at 90 MHz for most of the popular stand alone wideband FM signals, or with an optional 82 MHz stand alone oscillator for the Algerian approach to audio. At these frequencies, a relatively simple translator oscillator and double-balanced (or MOSFET) mixer will handle the chore.

THE CHALLENGE

As you can see, there are challenges here for equipment designers to develop additional variations to the now almost standard tunable audio subcarrier detection systems which most receivers offer. The rapid development of satellite systems for national use, using the Intelsat system worldwide, is leading to unique and perhaps novel techniques for the transmission of audio.

CHAPTER THREE

SATELLITE TELEPHONE SYSTEMS (SSB/FDM)

LOOK BUT DON'T TOUCH

(1) Section 605 of the Communications Act of 1934, as amended, generally warns you that non-broadcast-class communication signals are considered *"private" communications.* The 1934 act was designed to establish guidelines to prevent unauthorized "third parties" (i.e. anyone not associated directly with the "private communications") from setting out to intercept such private communications. Section 605 warns that any person not authorized to intercept such transmissions and who does so, either on purpose or accidentally, shall not divulge to any other party any of the following: (a) that such a transmission exists, (b) the content of the transmission "intercepted." In another section of the FCC rules, the FCC treats all microwave (i.e. satellite) communications in the 3.7 to 4.2 GHz band as "common carrier" transmissions, and thereafter confirms by definition that all common carrier transmissions are considered non-broadcast (as in private) transmissions. These rules were adopted at a time when the primary concern was with "eavesdropping" on private overseas and domestic telephone communications, and in studying the debate of the era, it would appear that the real concern was with employees of the telco/common carrier who might have access in their work routines to these circuits. The Congress clearly was worried primarily about common carrier employees eavesdropping. In the intervening 46 years the whole nature of communications has changed, including the use of some common carrier transmission frequencies for non-common carrier type signals. The satellite relay of television signals such as Chicago's WGN or radio signals such as Anchorage's KFQD does not fit the meaning or intent of the 1934 act, as far as "eavesdropping" on a "private" communications channel is concerned. No significant enforcement actions (by the FCC) have ever occurred in the 46 year history of Section 605, and the FCC is currently studying this antiquated section of law with an eye towards revising it to modern standards.

You've seen the signs. They warn you to keep your mitts off of someone else's property. (The sign may also mean the object is delicate or fragile, but we'll overlook that meaning here!). About 99% of what is on the satellites these days could, would, and may, in fact carry such a sign. In fact, of late some of the visual services have even begun to insert such warnings into their transmissions!

This is not a discussion of Section 605 (1). Rather, the footnote aside, we are going to ignore for the sake of science and the advancement of technology that there even is such a misunderstood, little practiced FCC "law" on the books. This is simply going to be a frank and open discussion of what is up there on all of those dozens, indeed hundreds, of 36 MHz wide transponders.

Satellite capacity is huge. We talk glibly about a transponder being 36 MHz wide, little realizing that that takes in everything from sound waves to well up into the lower portion of the VHF spectrum (i.e. 0 MHz to 36 MHz). Do you have any idea how many 10 kHz wide AM (as in broadcast radio) signals could be shoved into that much spectrum if somebody tried? Well, the answer is 3,600. That would be 3,600 evenly spaced AM broadcast stations, each covering all of North America, each with its own identity, all operating at once. Or, if you think a 10 kHz "channel" is very spectrum wasteful, how about 4 kHz single sideband channels? We could stack 9,000 of these in there all at the same time, and each (again) would have the ability to "talk to" anyplace in North America. These are impressive numbers.

And that's for one transponder. RCA and COMSTAR have 24 transponders and if one such satellite was utilized for AM radio, we'd have 86,400 "clear channel" continent wide, AM type broadcasting channels or 216,000, 4 kHz wide SSB channels to play with. Dare we suggest that even today there are 8 such satellite spots now being occupied by US birds, and that if all of these birds were turned over to either of our paper models there would be room for (are you ready!):

a) 691,200, 10 kHz wide AM stations, or,

b) 1,728,000, 4 kHz wide SSB channels.

These are such big numbers that we can ignore what would happen with another 8 or 9 orbit spots filled, or with the opening of yet a whole new set of transponders in the 11.7 to 12.2 GHz band. Obviously, there is a lot at stake out there in space. Big bucks ride on what happens, who makes it happen, and who it happens to.

This section of this manual is about some of the things you can find now, today, inside of these magic 36 MHz wide spectrums. Sure, you know all about TV and the wonderful world of R rated films, unending sports, non-stop religion, or Ted Turner's attempt to rewrite the news coverage business with dawn to dawn reporting. But you probably don't know where to tune in the current Kenai Flight Service Weather Forecasts from Anchorage, or National Public Radio, or several dozen Holidex (Holiday Inns of America) toll free reservation line curcuits, or a major computer company's inter-office 24-hour a day satellite relayed executive "tie-line"...and on and on and on. **Figure 3.1.**

1) The Perfect Spectrum — If our existing radio spectrum between 0 and 36 MHz was "perfectly used," it would operate like a single 36 MHz wide transponder on a geostationary/Clarke Orbit satellite.

How's that? Well, first of all the "satellite 36 MHz spectrum" is perfectly divided up (i.e. allocated) so that at precise (4 kHz) intervals, there is a new carrier signal capable of carrying or transmitting information from point "A" (the uplink) to any other point the satellite can "see." And since each carrier only carries a single set of "intelligence" (i.e. voice, data, etc.) there is never a worry about interference. Unlike the "real" 0–36 MHz spectrum where various groups, nations, users, and so on share and then fight over the "rights" to a particular frequency assignment, in "private" hands the satellite 36 MHz "spectrum" is constantly partitioned into discrete assignments and then those assignments are "protected," full time.

And unlike the "real" 0–36 MHz spectrum, this 0-36 MHz spectrum has all signals within it at essentially the same level of strength. That means that as a receiver is tuned across the spectrum, each carrier is perfectly separated, and there is either no modulation (voice or data, etc.) present because that particular frequency is not in use at that moment, or if there is modulation, it sounds exactly as loud as the carriers adjacent to it on both sides and hundreds of carrier assignments removed as well.

2) The Dedicated Spectrum — Unlike the "real" spectrum where there are constant battles for who uses or controls what particular frequency, the satellite spectrum with a 0–36 MHz transponder is "assigned" carrier by carrier to those users who have an immediate need for it at that point in time.

This means that when the assigned user has something to say (transmit), he says it; when he is listening to perhaps the return side of his two-way conversation, his side is essentially quiet. When there is no conversation going on, his assigned frequency is also "quiet." Each user knows that short of equipment failure, the circuit is his when he wants it, for as long as he needs it.

All of this takes a little getting used to, especially if you have had some experience tuning a radio around the real-world shortwave bands, or have ever tried to locate a "listenable AM broadcast station" on the car radio at dusk when you are in the middle of New Mexico.

HOW IT WORKS

There are two types of voice/data/signal channels on satellite these days. The most common type employs something called single sideband (SSB) as a modulation format. The technical name for this format is frequency division multiplex (abbreviated FDM-fm) which simply means that the baseband modulating frequency range (typically to 0 to 8 MHz) is treated like a spectrum of its own before the uplink transmitter gets a hold of the modulation. Individual carriers are each modulated with voice or data in the 300 to 3400 Hz (that's hertz!) region. The voice/data information is used to modulate a double balanced modulator where one of the two sidebands is eliminated with a filter. The remaining (sideband) signal is then applied to another carrier typically operating on some precise frequency between 64 and 108 kHz. See **Figure 3.2.** These sideband modulated carriers in the 64–108 kHz region are then multiplexed or mixed together in Groups of 12 to form a larger group of carriers called Supergroups. A Group has 12 carriers, a Supergroup has 60 carriers (consisting of five Groups). The Supergroup can be applied as a package directly to the baseband modulator of the uplink transmitter (which frequency modulates — fm's — the uplink transmitter), and they can in theory be placed anyplace from 0.0 MHz up to 10.75 MHz; that's the "spread" of the baseband modulating frequency (index). See **Figure 3.3.**

We'll see how all of this impacts on your receive terminal shortly; don't panic!

SATELLITE BUSINESS SYSTEMS

SATELLITE BUSINESS SYSTEMS
SBS SKYLINE
P.O. BOX 96502
CHICAGO IL 60693

BILL DATE NOV 27,

AMOUNT DUE $27.48

FOR ALL INQUIRIES
PLEASE CALL 800-368-6900
BETWEEN 9AM AND 9PM WEEKDAYS

SATELLITE BUSINESS SYSTEMS

PAGE 2

DATE	TIME	DESCRIPTION/ CITY CALLED	AREA-NUMBER	QTY/ MIN.	COST	AMOUNT
916	0930P	HOUSTON TX		58.6		7.62
918	0318P	REYNOLDSBGOH		43.3		6.06
918	0455P	REYNOLDSBGOH		2.1		.29
925	0812A	HAVANA FL		19.2		2.69
925	0801P	CULLOWHEE NC		7.5		1.05
925	1038P	CULLOWHEE NC		1.7		.24
1003	0843P	CULLOWHEE NC		1.0		.18
1008	0923A	HAVANA FL		33.8		4.73
1010	0633P	CULLOWHEE NC		18.1		3.26
	***	TOTAL CALLS VIA DALLAS				26.12
	***	TOTAL CALLS				26.12
FEDERAL EXCISE TAX OF			.78 ON	26.12		.78

Figure 3.1 *A phone bill showing charges of a satellite telephone system.*

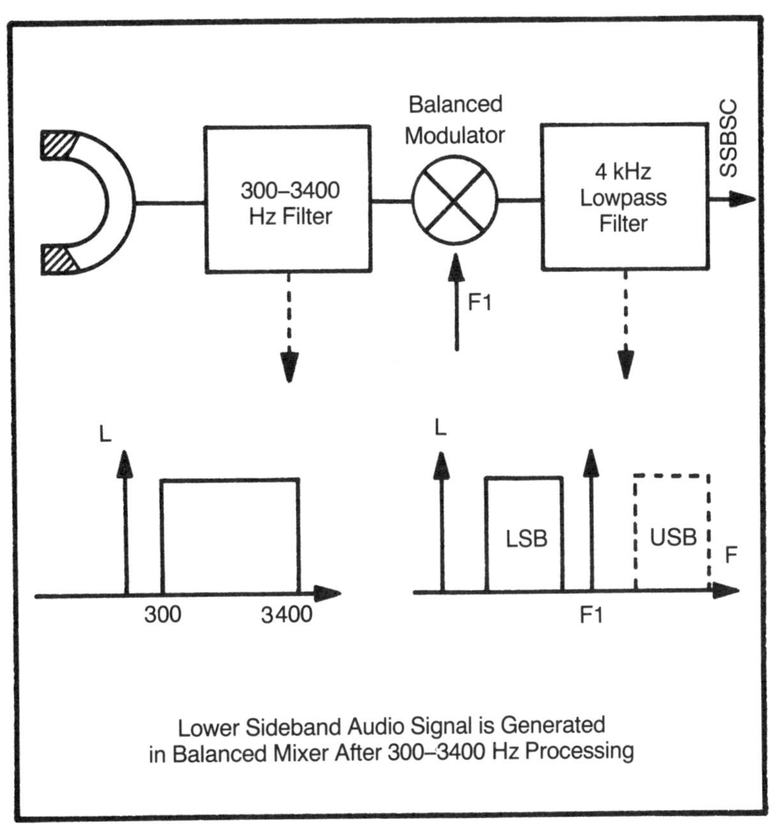

Figure 3.2 *Single side band system with one side band eliminated with filter.*

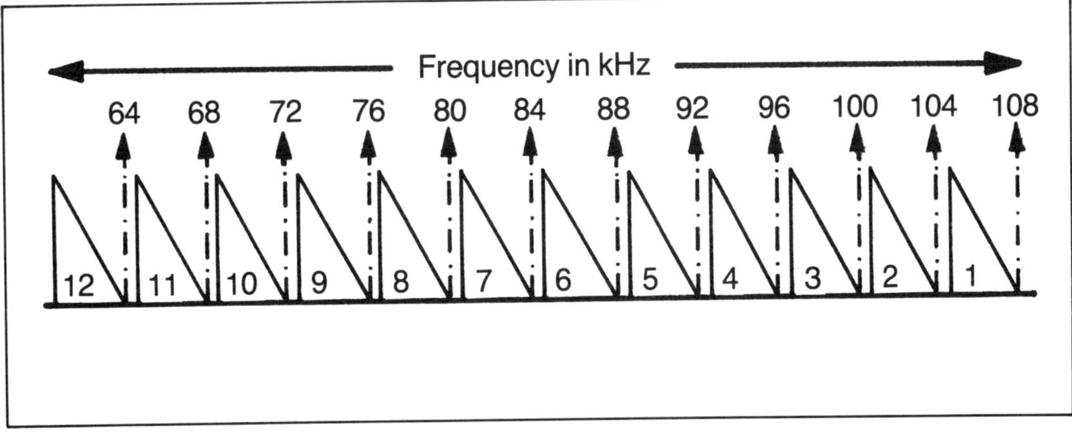

Figure 3.3 *Basic group of 12 "4 kHz" voice grade channels in the 64 kHz to 108 kHz range; each channel modulates the carrier separately. Type B group.*

Now because this "spectrum" can, and in fact is, carefully controlled by the uplink terminal operator, he wants to pack as many of these Supergroups into the basic 0.0 MHz (low end of "band") to 10.75 MHz (high end of "band") frequency range as possible. To do this, and to insure that equipment installed in Nigeria is compatible with equipment installed in Boise, a set of usually-followed "baseband allocations" have been adopted. Now these are not totally universal, but they come close. And when you are dealing with "allocations" within the same bird family, they are identical from transponder to transponder and bird to bird almost all (nothing is 100% in this life!) of the time.

A Supergroup of 60 carriers (each with its own modulation or information) is nice, but that is hardly large enough to accommodate the needs of the fast-growing satellite business. So there is an even larger group called a Mastergroup. This consists of five of the Supergroups, all mixed together at the baseband point before they are applied to the uplink transmitter. Let's see how all of this fits:

a) Group – consists of 12 carriers, each 4 kHz from the nearest carrier in the same Group; each capable of being modulated with audio or data with its own "baseband" frequency range of 300 to 3400 kHz. The basic carriers start off in the 64 (68, 72, etc.) to 108 kHz region where they are "modulated" with single sideband (lower sideband) techniques.

b) Supergroup – consists of five Groups all multiplexed together, plus five "multiplexing carrier frequencies." That means we actually have 13, not 12 carriers, per-Group in a Supergroup (see diagram here), and a total of 60 useable (for modulation) carriers in a Supergroup, plus five more that are part of the multiplexing process.

c) Mastergroup – consists of five Supergroups, again all multiplexed together, plus one additional multiplex-mixing carrier per Supergroup. This works out to 60 useable carriers per Supergroup times five, or 300 useable carriers per Mastergroup. See **Figure 3.4.**

Spectrum – A Group occupies 48 kHz of space; a Supergroup occupies 240 kHz of space while a Mastergroup takes up 1200 kHz of space. Got that? Well, it's not quite true! Actually, a Group does occupy 48 kHz and a Supergroup does occupy 240 kHz (that's 5 times 48), but a Mastergroup occupies 1232 kHz of space, not 1200. Why? Well, between each Supergroup in a Mastergroup, there is an 8 kHz buffer or guardband. We've diagrammed it here.

Now, why do they do all of this monkey business? To provide flexibility to the handling of "batches" of carriers, or Groups/Supergroups as they call these batches. Let's see why this is done.

AT THE UPLINK

At any given uplink site, there is typically more than one customer who is willing to pay the tariff rate for a dedicated (or non-dedicated, meaning less than full time) carrier-channel. In a metropolitan area such as New York City, there may well be several hundred users of the service and system. So at the New York City Toll Operations Center (TOC), all of the local customers come into the facility on either dedicated telco lines or by radio. Each is processed separately at this point of entry, and then the satellite TOC operator "looks at" each user and sees where he is going. If there are 12 users who want to be linked between New York City and Los Angeles, for example, the TOC operator selects each of these and "bundles" them into a Group. Then, he may have 52 that want service between New York City and Chicago; these he bundles into a Supergroup (by utilizing 5 Groups). He now has a set of "bundles," each with the same (metropolitan area) "point of origin" and each bundle is destined to end up at a separate point of delivery.

The various Groups and Supergroups (and perhaps even Mastergroups) are then all multiplexed (added together) at the TOC and sent on to the uplink site. They arrive there in a configuration where they can be applied directly to the FM (frequency modulating) system of the uplink transmitter. For sake of discussion, we'll say that within the uplink modulation baseband of 60 to 108 kHz we have a Group or bundle destined for Los Angeles, while in the 312 to 552 kHz region we have a Mastergroup headed for Chicago. See **Figure. 3.5.**

AT THE DOWNLINK

The downlink receive terminal in Los Angeles will receive everything on the transponder; however, it will (by instruction) pick-off only that (example) Group of

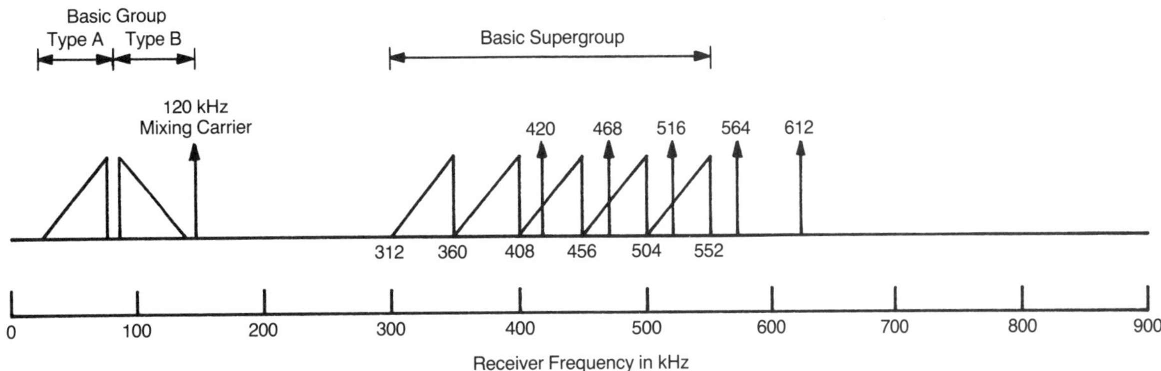

Figure 3.4 *Groups vs Super groups.*

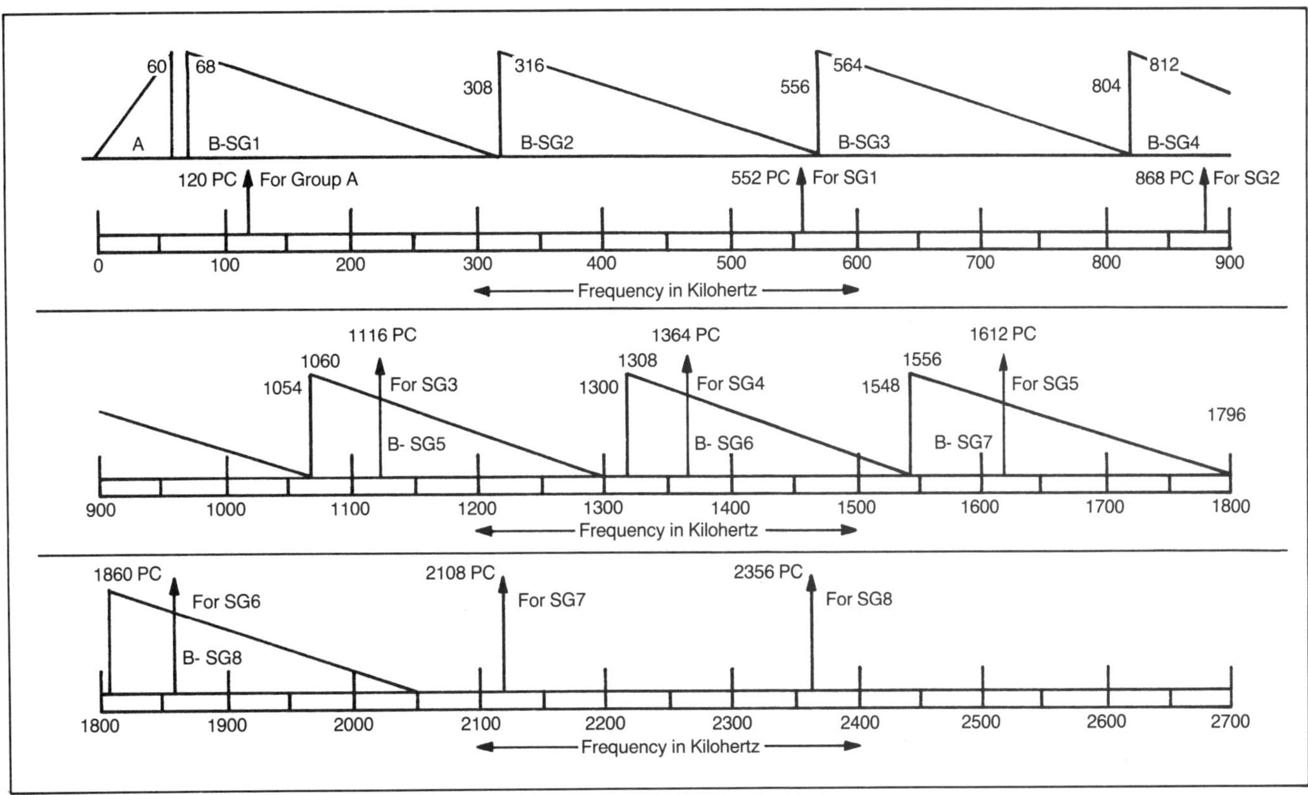

Figure 3.5 *Basic baseband spectrum in single sideband (SSB) frequency division multiplexing (FDM) on a typical satellite.*

signals between 60 and 108 kHz for local delivery. It will do this by demodulating all of the signals on the transponder to the baseband spectrum of 0.0 to 10.75 MHz, and then using block-conversion techniques extract the 60 to 108 kHz bundle. Since the downlink receive site is typically some distance outside of the metropolitan area itself (Los Angeles is located at a place called South Mountain for RCA), and because the destination customers are likely to be within the city itself, this "bundle" is then transferred to a terrestial microwave circuit (still as an 'intact Group') for delivery to LA proper. Once in LA at the TOC facility, the individual carriers within the Group are independently demodulated using a (lower) single sideband receiver, and the audio/data each contains is then plugged into a leased telco line or radio circuit that carries it to the final destination point.

Now in the case of our Chicago bound Supergroup, perhaps we end up with customers spread throughout the Chicago area. Let's say that by careful planning all of the satellite customers that have been bundled into the Supergroup—Group that is occupying spectrum in the 312 kHz portion of the Supergroup are in the Elmwood Park area of Chicago. At the Chicago receive site (which is out by Lake Geneva, in Wisconsin) it would be possible after demodulating from 4 GHz to baseband on that particular transponder to selectively take that frequency range of baseband (312 to 360 kHz) and send it as a Group on its own terrestrial microwave or radio link to a distribution site serving Elmwood Park. Then at Elmwood Park, the individual (12) carriers are demodulated to true-baseband (audio/data) and plugged into local loop telco leased lines for final delivery. At the same time, other Groups within the Supergroup received at the Lake Geneva receive-terminal can be sent either to individual local distribution centers or into downtown Chicago for re-distribution via leased lines.

The beauty of all this is the "bundling" or "packaging" that this approach allows. Rather than having to demodulate down to baseband audio (i.e. take the individual carriers to sound or data impulses) at each point, it only happens once; at the final distribution point just before it goes into a telco or radio link to the destination. Otherwise the signal is handled every step of the way at RF (radio frequencies), and these radio frequencies are bundled and batched and (block) converted around to get them into and out of the system as a group.

THE OTHER APPROACH

As noted in the opening paragraphs, there are two approaches to sending less-wide-than video signals via satellite. We have just described the FDM-fm system which is by far the most commonly employed. There is another system as well.

Not every would-be satellite communicator is so located that he has ready access to a TOC or uplink site. For such a user, just "getting into" the system or "receiving from" the system would involve a very healthy expenditure in terrestrial linking. A 100 mile private microwave hop for example could cost the would-be user $100,000, and that just gets him into the system!

As the number of uplink sites grow (and this is one of the fastest growing areas in satellites these days) sooner or later virtually everyone will be within a reasonable-cost terrestrial distance from an uplink. But sooner or later, maybe several decades. So to handle the interim problem a "remotely located" satellite communicator has to install his own uplink site. He may in truth share it with several others in the same area, but for now assume he is all alone. Now as a stand-alone, single-channel requirement user, he doesn't need all of that fancy multiplexing gear. What he needs is some way to transmit a single channel of voice or data into the bird, and in turn if he is actually engaging in a two-way operation, get a single channel back. This is called the SCPC (for Single Channel Per Carrier) technique and the name says it all. SCPC is not as spectrum effective as the FDM-fm approach; it cuts by around 50% the number of "circuits" that can be carried simultaneously through a single transponder, but it has the flexibility of being a stand-alone system. Using a relatively small uplink dish and relatively low power uplink transmitter (4.5 meter dish, perhaps as little as 30–100 watts of uplink power), the system can be installed quickly and put into operation in a few hours time.

The method of generating the signal (in the SSB format) is essentially the same with SCPC, and with some exceptions the signal will "sound" the same to a listener as an FDM-fm signal. The exceptions are important however:

1) Perhaps the user cannot cram all of his input modulation data into a 300–3400 kHz bandwidth. OK — the SCPC approach can allow wider bandwidths and they may or may not

be "recoverable" with a standard SSB receiver.

2) The uplink site may well not have the effective "power-per-channel bandwidth" as a full FDM-fm uplink site; this can cause the signal going through the satellite to appear to be considerably weaker than the FDM-fm signals also going through the same transponder.

In practice, the SCPC carriers may well be operating in a portion of the transponder all by themselves, although in theory they can be slotted in right next to the boundary of a Group or Supergroup or Mastergroup.

MULTIPLE INPUTS

One of the beauties of the satellite system is that it is not essential that all of the uplink signals going through a single transponder come from the same site. For example, it may be operationally desirable to load up a transponder so that there is a Group frequency band between 60 and 108 kHz originating at Lake Geneva, a Supergroup between 312 and 552 kHz originating at New York (Vernon Valley), a second Supergroup between 812 and 1052 originating at Atlanta, and so on. Each uplink site transmits to the satellite its own bundle and each receives the alternate bundles destined for its area.

In the same way, SCPC transmissions can be fitted in virtually anyplace that they will match the overall allocations scheme. Again, this is a very flexible (and quickly re-configured) system.

HOW TO RECEIVE ALL OF THIS

Now you understand the basics of how the signals are put together, and placed onto the bird. Now what happens at a downlink site?

First of all to recover these Groups, Supergroups and Mastergroups, we let our standard TVRO receiver do its thing. It receives the 3.7–4.2 GHz range, downconverts to some IF (such as 70 MHz) and then this IF is demodulated to baseband (i.e. a spectrum from 0 to some upper frequency limit less than 10.75 MHz). Your TVRO receiver may well have a "baseband output" jack, which up until now you never considered using for anything except as a source of video to plug into either a TV monitor or a TV re-modulator.

However, when you tune your TVRO receiver to one of the non-video channels, you are now demodulating (although it can't be seen on the TV screen, since it is not a video modulation format) all of the data that is there. Since you now understand that some or all of the spectrum between 0.0 and a maximum of 10.75 MHz may well have individual carriers (in Groups, etc.) present, naturally you want to know how to make these carriers "talk to you."

Remember that even though we have "demodulated" the 4 GHz signal with our TVRO receiver, the output of our TVRO receiver at this point is really still "modulated." Each of the carriers inside of the cable coming out of the receiver are still on discrete RF (frequencies) and each must be further demodulated from the typically single sideband format.

We do this by running a cable from the "video/demodulated/baseband" output on our receiver to a receiver that can tune the spectrum of interest. That spectrum is from 0.0 MHz up to as much as 10.75 MHz. If you are familiar with Ham or other communications equipment, this is starting to sound like a communications receiver. It is.

The receiver that will perform the final demodulation is a selectable sideband tunable receiver. There are dozens of different makes and models around, many of them on the used equipment market and selling for between $200 and $500. However, there are some warnings here:

1) Most receivers built for this application start off at 500 kHz (their "bottom tuning frequency/range") and tune upwards to 30 MHz. The upper tuning limit is no problem; you aren't going above 8 or 9 MHz most of the time anyhow, but, ideally you would like to tune down to at least 100 kHz.
2) Not all such receivers are SSB adapted; some (in fact many) were built years before SSB came along and they are AM (amplitude modulation) receivers with perhaps a BFO (beat frequency oscillator) thrown in for CW (continuous wave or Morse code) reception. You can use a BFO equipped receiver to tune in SSB or an older AM receiver, but you won't like doing it very much because it is unstable, tricky, and generally not much fun.
3) Some of the more recent receivers only cover the Ham radio bands (i.e. 1.8 to 2.0 MHz, 3.5 to 4.0 MHz, etc.). This will give you a "peak" at a portion of the spectrum, but not all of it.

THE IDEAL RECEIVER

In order to satisfactorily receive the telephone signals on the satellites, your general coverage, single sideband receiver must have the following features and capabilities:

1) The receiver must be stable frequency wise, after a short warm-up period.
2) The receiver must be able to receive both upper and lower sidebands.
3) Tuning must cover from 100 kHz to the 30 MHz range with ample sensitivity, and be able to tune 0 to 100 kHz with reduced sensitivity.
4) Receiver should have digital frequency readout that is accurate.

RECEIVERS USED

Some of the SSB general coverage receivers that have been successfully used are:

*Japan Radio Company's NRD-515 **Figure 3.6**
*Kenwood R-100 **Figure 3.7**
*Kenwood R-2000 **Figure 3.8**
*ICOM R-71A **Figure 3.9**

Plus, some of the new amateur transceivers that now carry full coverage receivers, such as the Kenwood TSR-430S, TRS-930S, etc.

PLEASE NOTE:

Many so-called shortwave receivers sold in discount stores and other retail outlets will not perform for this application; your best bet is to secure a receiver from a reliable amateur radio store or shortwave firm that stocks above quality receivers. The better names are JRC, Kenwood and ICOM.

THE HOOK-UP OF EQUIPMENT

1) *The baseband output of your satellite receiver* is fed to the antenna input of the quality SSB general receiver (.1 to 30 MHz) via a short length of 50 ohm RG-58U coax cable with the proper coax connectors on each end to the satellite receiver. TVRO receiver will no doubt have an "F" type fitting on the baseband output so you can obtain an adaptor from "F" fitting to PL-259 connector for this end of the coax cable, or your 50 ohm RG-58U can be fitted with a standard "F" connector at one end and a standard RF PL-259 fitting for the SSB receiver end. See **Figure 3.10** and **Figure 3.11.**

2) The output of the TVRO receiver must be baseband video that is not filtered. If when tuning the SSB receiver the signals are reduced in level after 4.2 MHz and then disappear, your TVRO receiver contains a filtering circuit that passes only signals in the 4.2 MHz video range. A simple remedy is to add a separate video output fitting that is wired ahead of the TVRO internal fitting. Many of the later TVRO receivers now come standard with this second video output. This is a simple matter for any installing dealer to do.

3) After the correct TVRO to SSB receiver hook-up is made, turn to a satellite that is known to be active with telephone traffic and choose a suitable transponder; at this point adjust the communications receiver to receive lower sideband. If you are capable of tuning 100/200 kHZ to 30 MHz with your receiver, just start dialing upwards from the "low" end. Every 4 kHz or so, you will hear a carrier. If it is not modulated at the moment, you'll hear an audio note in the 0–2 kHz region as you tune in and out of the carrier. When you run across one that is modulating at the moment, you'll hear a funny Donald Duck sounding voice, which will "clarify" (i.e. sound more normal) as you tune the signal in precisely.

4) Refer to the satellite activity section for satellites that are active with telephone traffic. See **Table 3-A.** After a little practice with tuning SSB, you can try going through the entire range again on the general coverage receiver in the upper sideband side of the signal, where you will find more traffic.

5) On some pure data and telephone transponders (non-video), you may hear something that sounds like a buzz saw, or something that sounds like musical chimes, or you may hear a telephone circuit ringing or a busy signal. On the active telephone channels, you will hear telephone conversations, some radio feeds, communications circuits between the satellite control operators, hotel and motel reservation sections, auto rental companies; at any given moment you could find between 600 to 1200 separate carriers in place! See **Figure 3.12.**

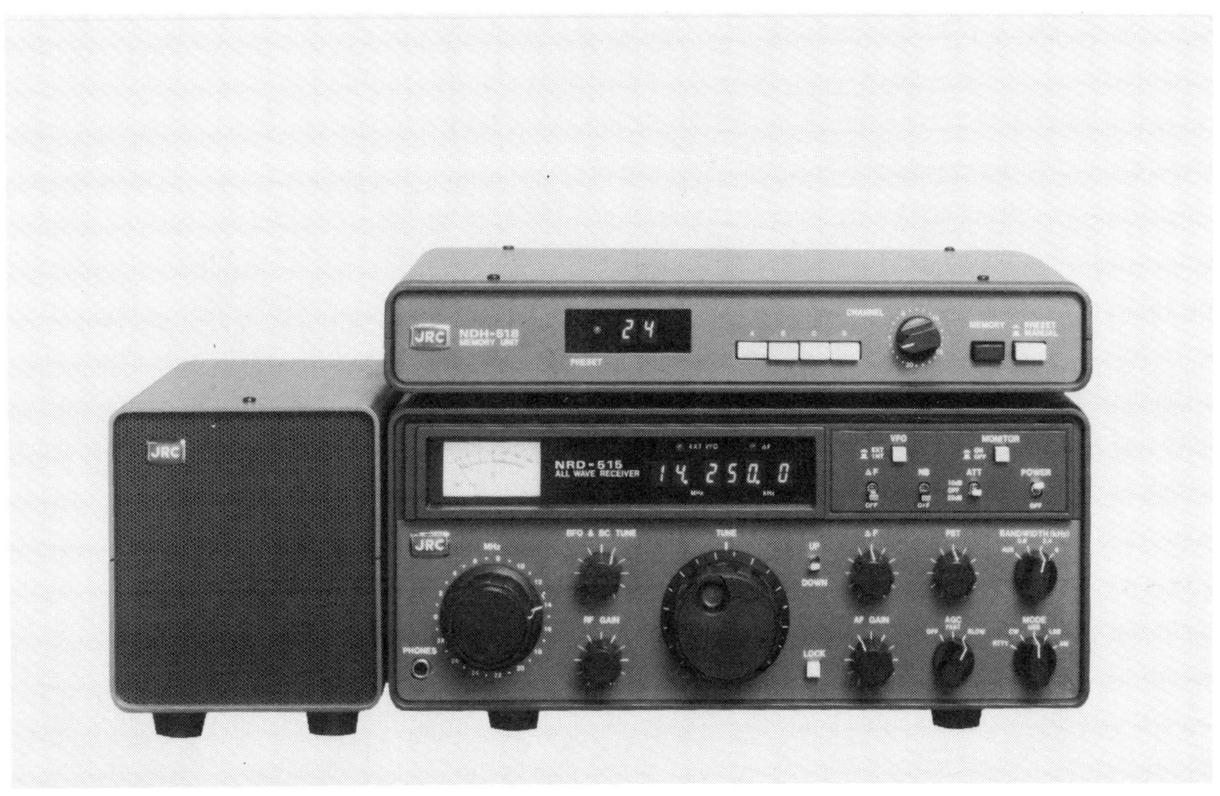

Figure 3.6 *JRC – NRD – 515 High quality SSB general coverage receiver .1 to 30 MHz coverage.*

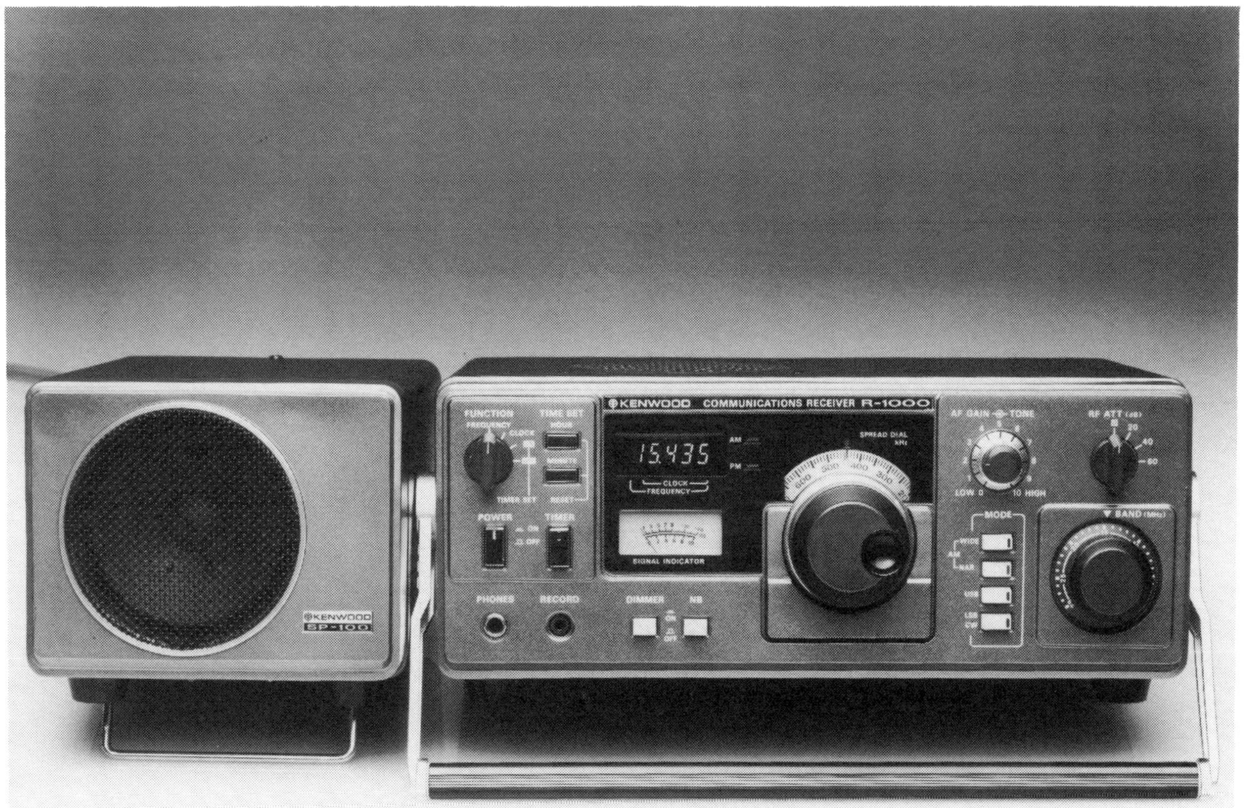

Figure 3.7 *Kenwood R–1000 general coverage SSB receiver.*

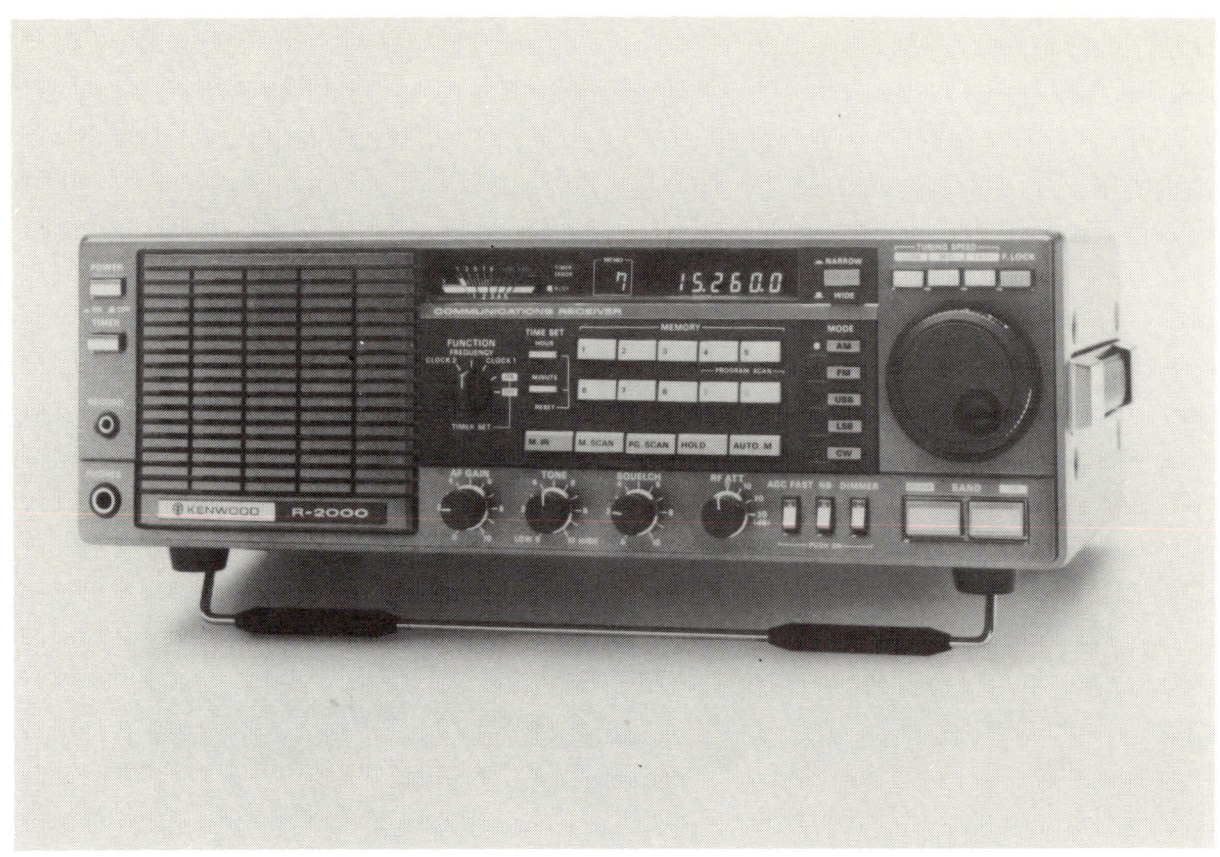

Figure 3.8 *Kenwood R-2000 general coverage SSB receiver.*

Figure 3.9 *ICOM R–71A general coverage SSB receiver.*

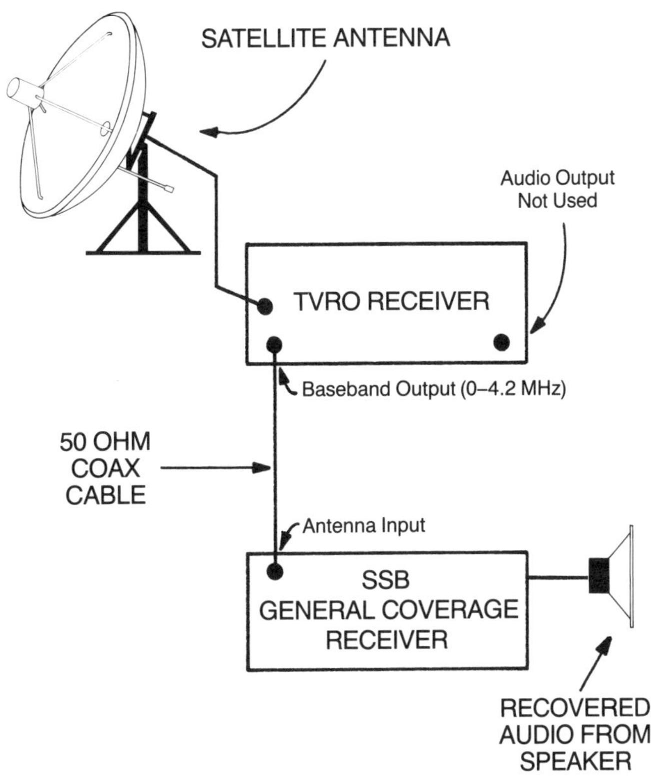

Figure 3.10 *Basic hook up of SSB receiver to TVRO for telephone reception.*

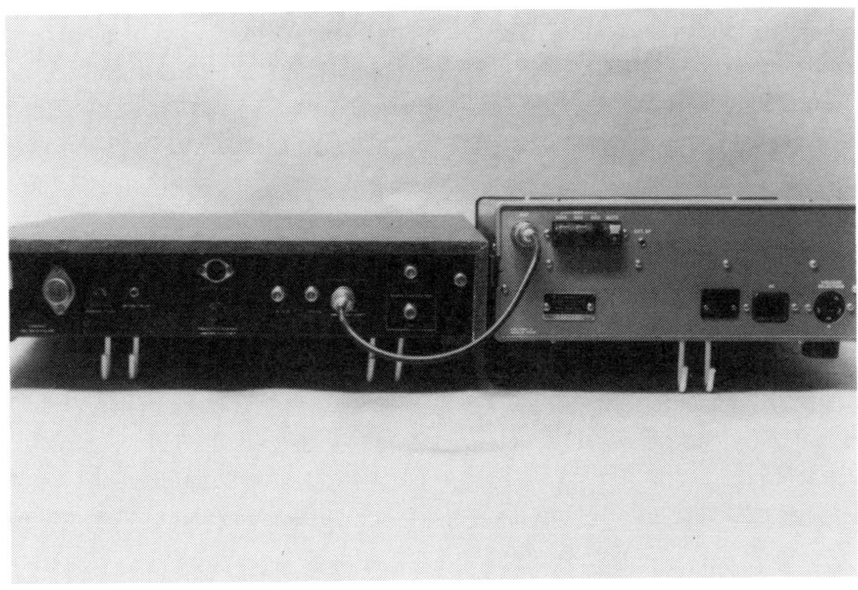

Figure 3.11 *Hook up showing COAX connection from TVRO receiver to SSB general coverage receiver.*

CONNECTION PROBLEMS

Your present TVRO receiver may be so designed that what comes out of the *"baseband" output* (where you plug in the communications receiver) is severely "filtered" above the normal baseband video frequency range of 0 to 4.2 or so MHz. What does this mean? Well, to keep the normal aural subcarriers associated with a video signal from getting into the video display circuits (i.e. either a monitor or RF video modulator), the receiver designer has built in a filtering circuit that passes only the portion of baseband that is going to contain video information. This cross-over point can occur anyplace from 4.2 to 4.5 MHz. That means that anything above this filtering point is going to get "chopped" out of the baseband output. That also means that any carriers modulated within the transponder above this filtering point are going to be severely attenuated or just eliminated.

What is the solution to this? If you run the system with the filtering in place, you won't copy the data above the cut off frequency very well, limiting you to the region below approximately 4.2 MHz. The best solution is to look for the point in your receiver at which you can connect an output line to drive the narrow band receiving equipment ahead of the filter. Some commercial receivers already have such an output jack available. After you locate the spot ahead of the filter, simply run a .001 disc ceramic capacitor from that point to a new (added) output jack and plug in the line to your communications receiver in the new jack.

The output of your TVRO receiver may be so strong that you will have to reduce the overall package gain of the communications receiver. The Kenwood R–1000 previously mentioned has a front panel switchable attenuator that lets you switch in 10 dB steps of attenuation. You'll know when there is too much "signal gain" ahead of the communications receiver; if it is equipped with an "S" (signal level) meter, the meter will indicate readings at the S9 mark or over, even when you are tuned between carriers (i.e. no signal being received).

As you tune between 100 kHz and the top end of the band, you may notice that suddenly a whole group of signals can't be tuned in (i.e. clarified). Don't panic—the uplink transmitter has simply multiplexed this Group or Supergroup into the circuit "upside down" and to tune them in, switch the communications receiver from lower sideband to upper sideband. We'll talk more about this shortly.

One of the advantages to receiving narrow-band signals is that they do not require as strong a satellite receiver signal for good quality reception. Why is that? Well, whenever the "bandwidth" (i.e. the amount of spectrum space required to transmit intelligence) is reduced, the whole system picks up additional sensitivity because you can employ a narrower (i.e. sharper) receiver to tune in the signals. For example, with a (big) 20 foot dish equipped with a 120 degree LNA and a standard receiver, these SSB signals typically measure signal-to-noise ratios in the 40 to 50 dB range. If you don't identify with numbers like this, trust us that these are good numbers indicating loud and clear reception. This is another way of saying that when you have a "marginal" TVRO picture (one with noise in it or on the verge of noise), the audio SSB signals will still be loud and clear on a system as described here. In fact, with a 6 foot dish equipped with a 120 degree LNA and the communications receiver system lash-up described here, you can still get "perfect" (i.e. noise free) audio or data reception.

In fact, the signals are so loud that when you are cross polarized (i.e. tuning in a vertically polarized transponder with the feed antenna oriented in a horizontal feed position), the signals are still good (if not perfect) copy on a 15 foot dish!

Receiver stability – an unstable TVRO receiver will be a problem. If your local oscillator stage (VCO) is not pretty stable and if your receiver AFC (automatic frequency control) system is not performing pretty well, you will shortly discover that you are "following" the narrow banded signals all over the dial with the communications receiver. The solution may not be simple, but it should be obvious—stabilize the receiver!

If your receiver has an AGC (automatic gain control) system, disable it before using the communications receiver for what follows:

a) Your TVRO receiver seldom has an accurate signal level meter indicator; one that tells you when you are making small improvements (or degradation) in the aiming of TVRO antenna. The communications receiver will typically have a good quality "S" (signal level) meter on it.

b) Having disabled the TVRO receiver AGC (if there is one), tune in an unmodulated carrier with the communications receiver (using the RF gain control or a switchable front end pad such as the R–1000 employs), so that the carrier you have tuned in reads around "S9," typically mid-scale, or lower.

c) Now, as you adjust the TVRO antenna azimuth and elevation, watch the "S" meter on the communications receiver. The rule of thumb is that for each 1 "S" unit you gain or lose on the communications receiver "S meter," you have just picked up or lost 1.5 dB C + N to N (carrier plus noise to noise) at 4 GHz. In other words, since a "S" unit is usually 6 dB ("S meters" are typically calibrated so that all of the "S" marks from 0 to 9 are 6 dB steps, while above "S9" the meter is calibrated in 5 or 10 dB steps), 6 dB improvement in a 2.4 kHz bandwidth works out to around 1.5 dB change in the 4 GHz C + N to N ratio. As long as the carrier you are tuned into stays unmodulated, it will be an excellent "test signal" for fine tweeking on your TVRO dish.

In other words, perhaps the only real "justification" you need to add a communications receiver to your TVRO installation is the fact that you then have a very accurate "metering system" with which to align your antenna to the various satellites!

Now when you are working with very weak signals as when you are getting a system operational, or perhaps you are trying to make INTELSAT's feed for Brasil look good, you will find you can tell if things are better or worse much quicker and with much greater accuracy with a communications receiver tuned to an unmodulated carrier on a transponder. I have yet to find any satellite that doesn't have at least one audible (on a communications receiver) carrier on it if it is in the TV relay business. These carriers are there for many purposes including providing telephone intercom links for the people operating the satellites. Since people do not communicate all the time, when they aren't talking the steady unmodulated carrier is there for you to use in your measurements.

PLEASE NOTE:

You must be sure you are getting pure unfiltered baseband video—that is, the entire spectrum of the video and audio where all hookups call for baseband video and baseband audio, sometimes called unfiltered outputs. To install the proper output, see proceeding section or check with your dealer or service center to have this simple installation done for you.

If you do not seem to be getting proper results from any of the outlined hookups, LOOK TO THIS AREA FIRST! Make sure you have the full baseband output on video if you require video connection, or full range unfiltered audio if you require audio connection. Almost every case I know of where lack of reception was a problem, the cause was traced to the output of the TVRO receiver *not furnishing* true baseband video or audio out.

At this time there are only two TVRO receivers on the market that have the two necessary unfiltered baseband video/audio outputs. They are as follows:

1) USS/MASPRO SR-2 from United Satellite Systems, St. Hilarire, MN, 56754.
2) MACOM M-1 receiver from Macom, Box 1729, Hickory, NC, 28603

These same baseband requirements will be a requirement for the use of TVRO scrambling decoders, which are used by the major program suppliers, HBO, Cinemax and others. The present system by Linkabit Corporation produces the video Cipher III unit, a standalone TVRO decoder for the home market.

We expect all of the new generation of TVRO receivers will come equipped with these outputs as standard now that the programmers have progressed into scrambling.

WHAT IS MULTIPLEXING?

As satellite space becomes more and more crowded, new and innovative techniques of optimizing and expanding the amount of information that can be transmitted through a transponder will be required. One effective way of increasing the use of the system is to expand service by means of adding subcarriers and then providing multiple audio channels on the subcarriers.

The term "multiplex" designates a system in which two or more individual signals can be transmitted over a single carrier or other transmission medium. Multiplexing was first used in the early 1900's, when the telephone systems were faced with huge expansion of their existing system. Instead of running thousands of miles of additional cable to handle this needed expansion, a new technique was developed to allow more than one channel or voice use per wire which is now called frequency division multiplex (FDM). Since then these companies have refined and improved this technique to many thousands of voice channels over a single wire. Also using FDM up to 2700 voice channels can be transmitted over terrestrial microwave links on a single carrier.

These narrow band signals are now used in many forms on satellite transmission today and will increase many times in the coming years. The telephone voice channels are only one use of multiplexing of many signals on a given carrier. Multiplexing techniques and multiplexing will be dealt with in other chapters in greater detail, and combined with other transmission systems.

WHAT WILL WE HEAR?

Many of the telephone circuits you will hear via the TVRO receiver and general coverage single sideband receiver will be one side of the telephone conversation; the other side or other party will be transmitted on another set of carriers; these could be on the next set of signals or they could be on another transponder on the same satellite. We have heard many duplex, i.e. both sides of the telephone circuit on the same carrier or same signal.

The present plan used by most systems seems to be groups or packages of circuits all going toward one general direction, with another group going the other way. A package or large group of telephone conversations would be from the East Coast of the United States going to the West Coast, with another set or group going one way from Chicago to the West Coast and another set from the West Coast going back to Chicago to form the two-sided telephone complete conversation.

As covered earlier, the telephone circuits use the multiplex system to combine many thousands of telephone-audio channels on a single satellite transponder; this allows the companies to arrange many circuits throughout the entire country.

In conclusion, as with all private transmissions, these should not be used in any way for personal gain. This entire area is dealt with in the Communications Act of 1934, section 605 and are non-broadcast class communications signals, considered "private" communications. Review this act on the title page in front of the book, as well as in the general information section in back of the book.

LOCATION OF SATELLITE TELEPHONE SYSTEMS

The locations of most of the telephone systems seem to have stabilized at this date. Listed on Table 3-B are some of the established locations. As you progress with your receiving equipment, spend some time exploring other satellites and new satellites for telephone and data traffic; we are sure you will be able to find many more as new channels and new satellites appear in the Clarke belt.

Please make sure your polarization is correct for the listed satellite, and the listed transponder. Use your signal strength meter on the TVRO receiver as a first step, followed by a final peaking using your SSB general coverage receiver's "S" meter.

The SSB receiver makes a great signal strength measuring device as its sensitivity is much higher than the TVRO receivers signal strength meter.

TRANSPONDER FREQUENCY/POLARIZATION

FREQUENCY	TRANSPONDER/POLARIZATION SCHEME			
	A	B	C	D
3720	1 V	1 H	1 H	
3740	2 H	2 V		1 V
3760	3 V	3 H	2 H	
3780	4 H	4 V		2 V
3800	5 V	5 H	3 H	
3820	6 H	6 V		3 V
3840	7 V	7 H	4 H	
3860	8 H	8 V		4 V
3800	9 V	9 H	5 H	
3900	10H	10V		5 V
3920	11V	11H	6 H	
3940	12H	12V		6 V
3960	13V	13H	7 H	
3980	14H	14V		7 V
4000	15V	15H	8 H	
4020	16H	16V		8 V
4040	17V	17H	9 H	
4060	18H	18V		9 V
4080	19V	19H	10H	
4100	20H	20V		10V
4120	21V	21H	11H	
4140	22H	22V		11V
4160	23V	23H	12H	
4180	24H	24V		12V

SATELLITE TELEPHONE SYSTEMS (SBB/FDM)
GEOSYNCHRONOUS SATELLITES WITH TELCO TRAFFIC

SATELLITE	WEST LONG.	TRANSPONDER/POLARIZATION SCHEME
SATCOM	143.0	A
SATCOM IR	135.0	A
COMSTAR IV	127.0	A
WESTAR IV	99.0	B
TELSTAR IIIA	96.0	A
COMSTAR III	87.0	A
WESTAR II	79.0	B
GALAXY II	74.0	A
SATCOM IIR	72.0	A

Table 3A *Telco Locations*

SATELLITE TELEPHONE SYSTEMS (SSB/FDM)
LOCATION OF SATELLITE TELEPHONE SYSTEMS

SATELLITE	LOCATION	TELCO ON TRANSPONDERS
SATCOM V	143.0	3–5–7–11–17–23
SATCOM IR	139.0	5–7–9–10–17–21
COMSTAR IV	127.0	1–3–4–6–7–15–16–19–22–23
WESTAR IV	99.0	14–20–24 *
TELSTAR IIIA	96.0	*
COMSTAR III	87.0	2–5–6–7–9–14–15–16–18–20–21–22–23
WESTAR II	79.0	1–4–5–8–9
GALAXY II	74.0	* 12 TRANSPONDERS TO MCI
COMSTAR 01/02	76.0	** ALL TRANSPONDERS IN TELCO USE
SATCOM IIR	72.0	3–4–7–19–21–22–23–24

*SEARCH ALL TRANSPONDERS ON THIS SATELLITE
**NEARING END OF ACTIVE LIFE-EMERGENCY SPARES

Table 3B

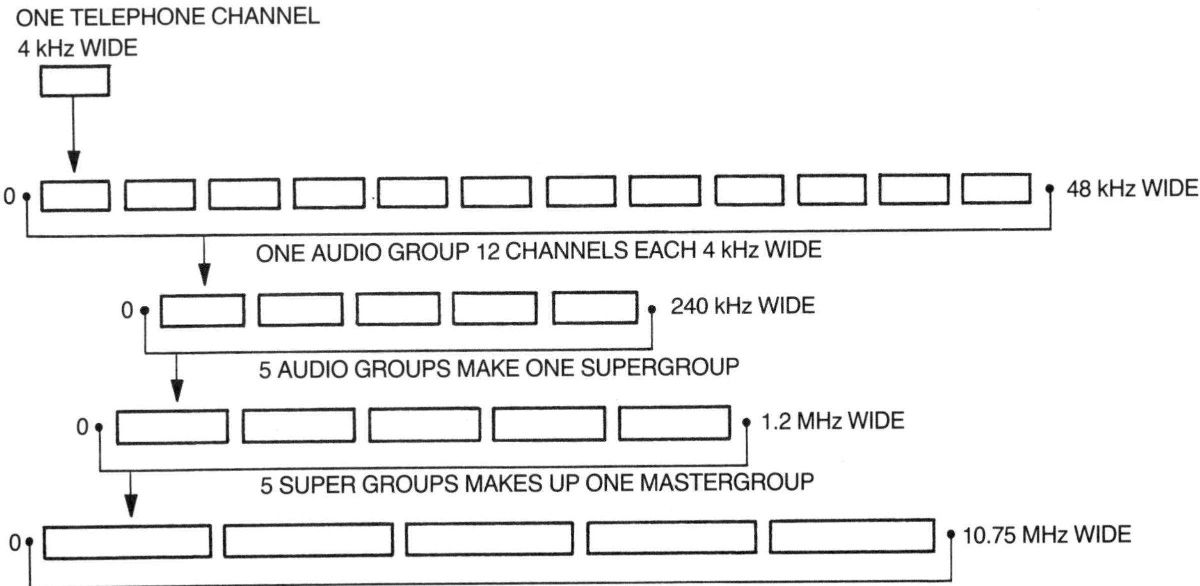

Figure 3.12 *Basic multiplexing telephone system used on many satellite systems. A single side band multiplexing mode.*

CHAPTER FOUR
SINGLE CHANNEL PER CARRIER (FM/SCPC) (SSB/SCPC)

FM–SCPC WHAT IS IT?

Several years ago, when satellite TV was new and everyone was learning, there was a flurry of interest in non-video satellite services. In several early TVRO manuals circulated in the 1980/81 era, entire chapters were devoted to the systems required to bring down from the birds network radio services, press services (UPI and AP), teletype® and a host of other "narrow band" links.

In the intervening years, the intense interest in video has overshadowed what was a growing interest in non-motion, non-video services and information concerning the available services became more and more difficult to obtain. Today very little is known about the current state of those services, or the type of 1984-type hardware required to bring those services into a private location.

There has always been two separate methods of transmitting narrow band material via satellite. In 1980, one of these formats (SSB/SCPC) predominated while another, perhaps technically better and technically more complicated, FM/SCPC, was a newcomer. In the interim years, the FM system has taken hold and most of the high quality audio and data links in use today, on both domestic and Intelsat satellites, rely on this frequency modulation system.

Let us first define the market for such services, since as system dealers we must always equate what is available to "who we can sell it to" as a business activity. Then we will take a preliminary look at the equipment required, and where you go to find that equipment.

RADIO NETWORKING

The development of radio networks in the late 20's and early 30's provided the foundation for the modern-day broadcasting industry we have in North America. And radio networking depended upon the ability to interconnect two or more radio broadcasting stations so that each could release or transmit the same program at the same time. The economies of networking are considerable; networking allows a massive, national or international audience to participate in the same event or program at the same time. It also attracts national advertisers, who in turn can build their own product distribution programs around their abilities to reach millions of homes instantaneously.

Radio networking depended for years on the Bell Telephone Company landline network for interconnection. In heavily populated areas, coaxial or other types of cables carried radio network signals from town to town and city to city, dropping off the "feeds" at each community where there was a local broadcast affiliate. The broadcaster, in turn, would dedicate one (or more) audio "lines" on his control console to the directly-connected network service line, and the on-air engineer would, on cue or at an appointed time, simply fade down the audio from the local programming and "bring up" the audio from the network programming. In this way, local news and programs were integrated with national news and programs to make up the full broadcast day.

In more remote areas where the population density did not justify the dedication of full time "wire lines" to radio network program carriage, stations were dependent upon either a microwave interconnection, for their network "feeds," or alternately, on picking up the over-the-air broadcast from another station affiliated with the same network, using an "off-air" receiver. In either case, the end result was the same; the audio from the feed terminated on the control room console and the operator on duty could switch to the network feed as the radio station's schedule dictated.

It is and has been standard practice for the network operators (Mutual, ABC, et. al.) to pick up the "tab" for the interconnection service. This means that the networks have always been large users of Bell circuits, and their bill to use those circuits has been substantial. Bell likes to plan circuits and systems a decade or more ahead, and when Mutual (for example) went to Bell in 1960 with their network service plans, they were talking not about the network service plans for 1960, but rather for 1970. Getting Bell to move "faster" was not impossible, only expensive. The Bell network, as wonderful and reliable as it is, does not lend itself to inexpensive, *rapid*-change, so when a special event came along that required monstrous re-routing of network services, there were monstrous charges associated with the changes.

Bell also has always been "network quality line" limited. While it is possible to send human speech ("voice grade") material over a standard telephone line, the standard telephone lines lack several important parameters which are required for *quality* radio networking. Bandwidth is one of those limitations; the standard Bell telephone circuit is typically capable of "passing" audio only in the 50 Hz to 3,000 Hz region with good clarity. Most musical programs contain passages which far exceed this relatively narrow bandwidth; a 15,000 Hz bandwidth is more in keeping with what we normally call "high fidelity" audio. To handle this, Bell circuits have to be specially "equalized" using equipment which is dedicated to broadcast circuit use. Even human speech suffers in a *normal* telephone interconnect; you can hear the difference yourself when a radio announcer is interviewing someone on the telephone and his voice is transmitted to you through his in-studio microphone, while the other voice is coupled out of the telephone. The telephone source voice has a "tinny" sound, when compared directly to the in-studio microphone voice. Thanks to modern electronic audio processing equipment, some of this can be artificially corrected today, but the correction is limited to handling the human voice range and does not extend to the musical instrument range. See **Figure 4.0.**

It was a natural, then, for radio networks to look with favor on the establishment of satellite linked interconnection systems. First of all, the circuits compared very favorably in cost to Bell landline circuit rental, if you "amortized" the costs over 5 to 7 years. Second of all, satellite links could be (and were) designed from the ground up to be capable of handling all of the audio range required (20 Hz to 20,000 Hz is not an unusual specification). Next, whereas most radio networks rely primarily on one "circuit" for most of their transmission needs, there are occasions when a second or third circuit would give them the flexibility they need to do a better job. Satellite links can be "stacked" with two or more channels for a minimal increase in cost. This gives the networks an expansion capability, on short notice, which they never had with terrestrial landline circuits. And finally, while landline networking was barely able to handle "high fidelity" monaural audio (with some difficulty) using satellite circuits, the networks could stack a pair of separate "channels" and send stereo service just as easily as they were sending mono over on the landlines.

The Mutual Radio Network was the first of the "big four" radio networks to invest heavily in the satellite linking project. Mutual began in 1978 with a proposal to equip its affiliates with modest-sized dishes (ten feet in diameter). *At the time* of the proposal, the FCC was *not* allowing *anyone* to use 10 foot dishes for anything; you still had to have a satellite-receiving-terminal "license" and the Commission's view of small dishes was that they did not meet their technical requirements. It was at about the same time that cable system operators and the first wave of private terminal owners were trying out dishes in the 10 and 12 foot region, and it took the combined "weight" of all of these factions to convince the FCC that smaller dishes were "OK." Mutual played a substantial role in our "non-license" status we enjoy for home TVROs today, simply because they were aggressively pushing the FCC to approve dishes in the ten foot region.

Since 1978, radio networking has grown up dramatically. Some of the networks supply their affiliates with small "ARO" (Audio/Receive Only) terminals. Others provide the satellite service, but require the affiliates to purchase their own terminals. Major suppliers of ARO terminals include Scientific Atlanta, Microdyne/AFC, California Microwave, and others.

The basic ARO terminal consists of an antenna plus LNA, a down converter, and a special "ARO" terminal demodulator. The system, up to the demodulator, looks like and in fact works just like your typical ten foot region home TVRO. Very few of the ARO terminal systems are equipped with motor drives, however, since the radio stations (and others) using these services don't move around the sky looking for feeds, the feeds are dedicated to specific transponders on specific birds full time.

The ARO Demodulator is an interesting piece of equipment. See **Figure 4.1.** There are quasi-standards

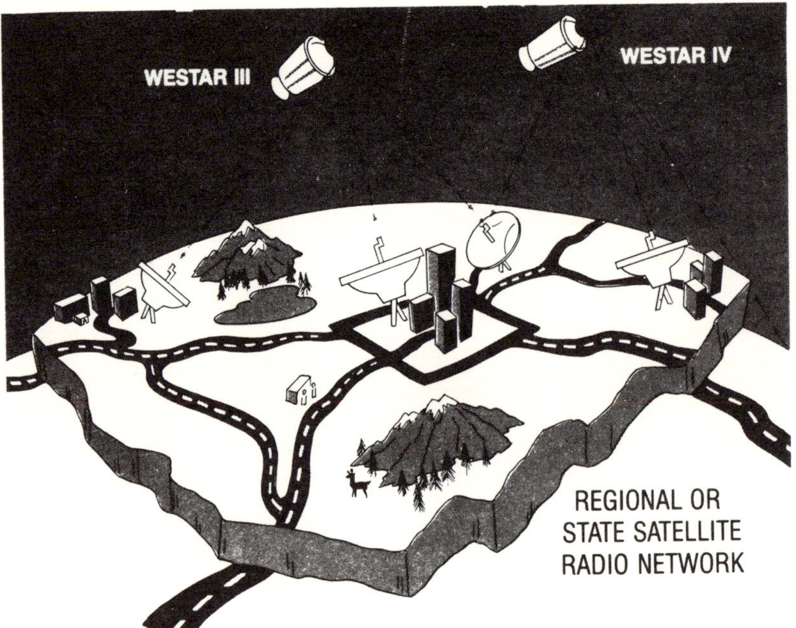

Figure 4.0 *Radio networking via satellite.*
Courtesy Comtech Corp.

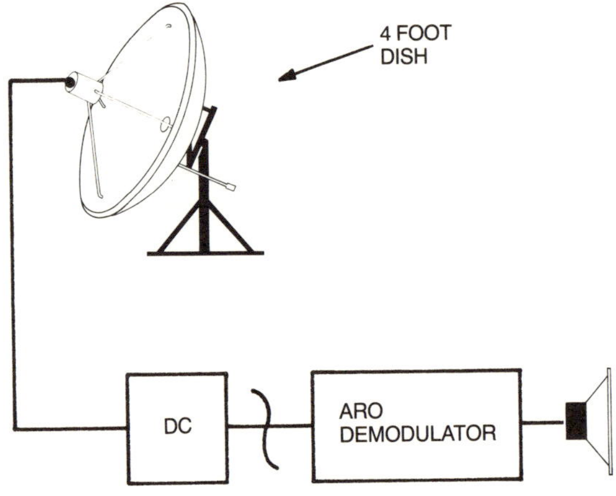

Figure 4.1 *Dedicated 4' ARO dish offers quality FM/SCPC service with small dish.*

in this field, and the most frequently utilized standards place the individual audio or data signals inside of a transponder (such as TR3 of Westar 4) 200 kHz apart. The individual "channels" occupy a bandwidth of 60 kHz. How do these numbers relate to TVRO/subcarrier audio? See **Figure 4.2**

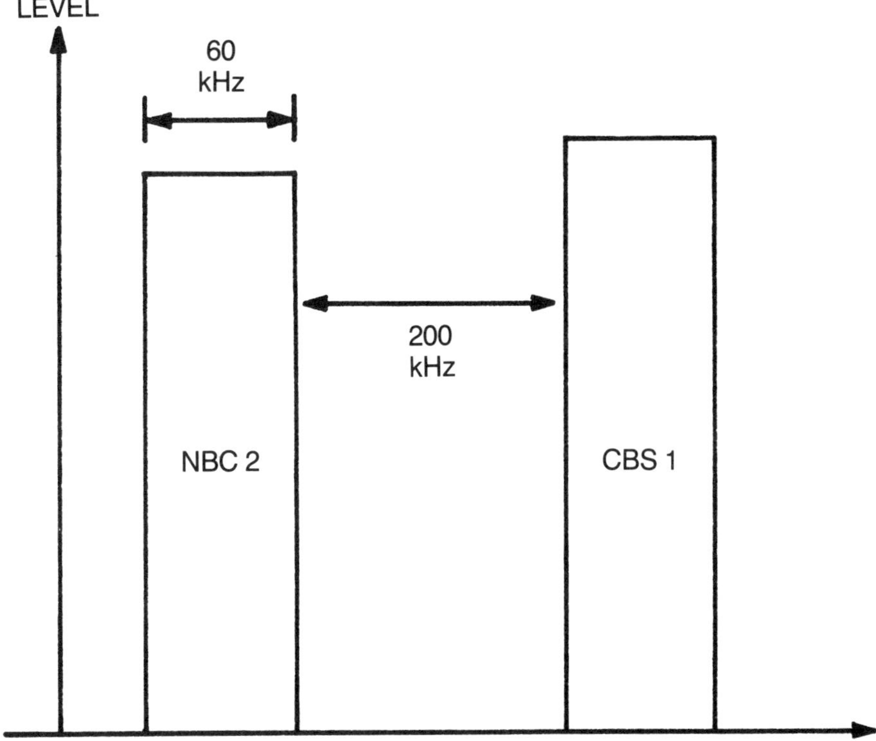

Figure 4.2 *FM/SCPC services can be found spaced, typically, 200 kHz apart (center to center) with each 60 kHz wide.*

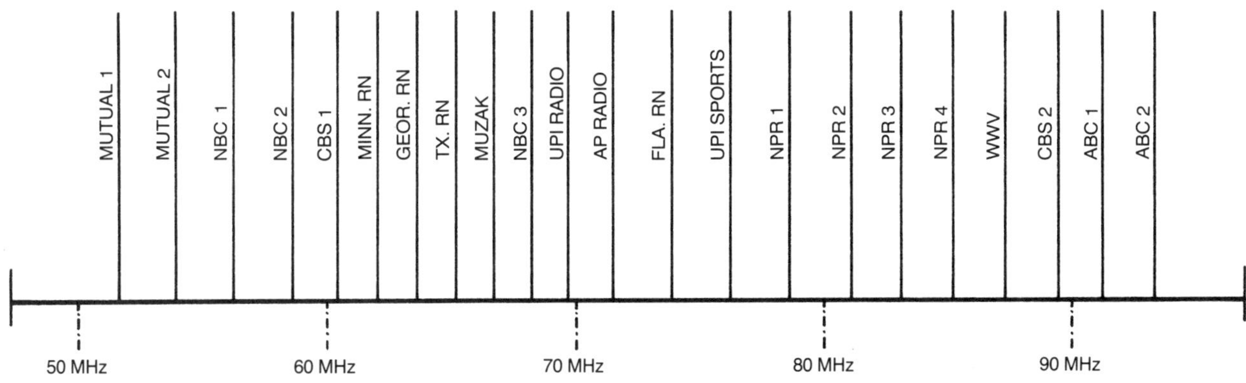

Figure 4.3 *Typical (not actual) FM/SCPC single transponder service assignments.*

First of all, if the transponder has no video, then the entire transponder can be divided up into a sub-world all to itself. The satellite operator (i.e., Western Union) can look at his 36/40 MHz of transponder space, and he can decide which service he wishes to assign to what frequency within that transponder. In effect, whereas the FCC does the "frequency-allocating" for terrestrial broadcast services, the satellite operator becomes "the FCC" for the on-transponder services. The services are assigned space or a specific operating (uplink) frequency. So far this sounds pretty much like United Video's carving up of the "subcarrier spectrum region" on WGN's TR3 transponder of F3R. The subcarriers on F3R differ considerably, however, from the "stand alone" FM/SCPC carriers found on TR3 of Westar 4. The F3R subcarriers are actually *married to* the video carrier, and they in essence run along with the video "free of charge." The uplink operator has to be concerned only that he keeps his subcarrier sufficiently spread out so that they do not "cross-talk" to one another, and to insure that as he adds new subcarriers, he does not "rob" too much power away from the video carrier (each audio subcarrier added to a video carrier reduces the effective power of the video carrier).

The FM/SCPC service allows each of the network services operating with a transponder the ability to uplink their own signal. In effect, if you had 20 different radio network signals on a transponder, each independent in content from the other, you could also have 20 separate uplinks *all sending signals* to the satellite. Those uplinks could be all in one geographic region (i.e., New York City), or they could be spread (as they typically are) from coast to coast. So *that* is a difference. See **Figure 4.3.** Another difference is that the bandwidth of the FM/SCPC services is far narrower than the subcarrier audio service; rather than being as much as 300 kHz wide (i.e., Disney), the FM/SCPC carriers are pretty much standardized domestically to a 60 kHz bandwidth. This simply means that if you try to use a receiver designed for one service to tune in the other format, you won't recover proper (or any) audio.

However, the major difference between subcarrier techniques and stand-alone SCPC techniques; *with subcarrier operation,* is that first you have to receive (and demodulate) the video carrier. Once you have the video demodulated, you *now* can go to a "second demodulator" and extract the audio from the video signal. In effect, inside your TVRO video receiver you have *two separate* receivers, one that makes pictures, and having made pictures, another that makes sound. The FM/SCPC format is more akin to your standard FM broadcast service, as each individual station "stands alone" and you tune it in with a master tuning dial. In fact, the only significant operational difference between tuning in a dial full of FM broadcast stations on a home tuner, and tuning in a transponder full of FM/SCPC signals is that you are tuning the satellite frequency band (indirectly) with the "ARO" receiver, rather than the FM broadcast band.

This tells us that what we need to receive "ARO" service is a receiver which operates like a standard FM tuner/receiver, except that it does it in the "satellite TV band;" and with the bandwidth system which the typical radio networking link uses. And that brings us back to the typical "ARO" terminal.

THE SYSTEM

There are two major differences between an FM broadcast band tuner/amplifier, and an "ARO" receiver/tuner.

1) The FM broadcast band is 88–108 MHz (in North America); the satellite band is 3,700 to 4,200 MHz;
2) *Some* (but not all) of the satellite FM services (sFMs) employ the same 30 Hz "energy dispersal wavetorm," or dither, which *most* of the video services use. This is an FCC requirement (although not all follow it); a system to reduce the likelihood that a satellite signal in the 3.7 to 4.2 GHz frequency range will interfere with a terrestrial (Bell) link in the same frequency range.

The smart approach is to shift the satellite frequency range to a much lower frequency range, just as we do with TVROs. This requires a down converter (after the LNA). There are two possible approaches here, just as with TV:

1) *Shift the full band* (3.7 to 4.2 GHz) to a lower "block of bandwidth," say 270 to 770 MHz (i.e., AVCOM, Scientific Atlanta, some of the newer Microdyne receivers). Getting the satellite "band" down to a lower frequency range makes it far easier to "transport" the satellite signals from the antenna/dish (where the down converter is located) "inside" to the demodulator portion.

2) *Shift just one transponder* at a time from its assigned frequency range (say TR3, 3840 to 3880 MHz) to a lower "IF" range such as 50-90 MHz (70 MHz center frequency). See **Figure 4.4.**

Once the signal is shifted in frequency to a lower intermediate frequency (i.e., IF) we can then build a receiver which "tunes" that IF region, say 270 to 770 MHz, or, 50 to 90 MHz. It turns out that in a block down conversion system (such as 270–770 MHz IF), we are still asking a lot of receiver design engineers to turn that relatively high (UHF) frequency range signal(s) into video, or audio. So the BDC is usually an intermediate step, after the 270–770 MHz IF which leaves the antenna mounted down converter, we have a second "conversion" of frequency inside of the demodulator. The 270–770 range then becomes 70 MHz where not the *full* band *but one transponder at at time* is processed. So whether we start out leaving the outdoor down converter at 70 MHz (so-called single conversion system); or, we first block convert to 270–770 MHz, and then after transporting the "block" inside, reconvert a second time to 70 MHz, we usually end up at the "standard" 70 MHz IF anyhow. See **Figure 4.5.**

So here we are building an FM "tuner" which will cover the frequency range of 50–90 MHz, nominally. We process the audio-loaded transponder, with perhaps several dozen separate and distinct radio network type signals, to a standard 70 MHz IF (which is actually an IF that centers on 70 MHz, but which typically covers from 50/55 to 90/95 MHz). Then we build our FM tuner, following the specifications adopted for sFMs. This is your basic professional grade FM/SCPC receiver.

This brings us smack up against the first substantial problem with the system: *frequency stability.* Early TVRO receivers had something called "drift." This meant that the user had to touch-up the channel selector tuning control every few minutes, or 30 minutes, to keep the signal "tuned in." A microwave receiver has many demands placed on it; staying tuned on the right signal, all of the time, is one of those demands. A circuit called "AFC" (automatic frequency control) was the answer. By using the television video signal received by the demodulator as a "reference," the AFC system *constantly monitors* the frequency-centering of the received signal. If the centering moves because some of the circuits in the receiver are not totally "stable," the AFC corrects the tuning, automatically. That ended the constant across-the-dial chasing most of the early home TVRO receivers had. Unfortunately, this technique requires that the receiver have a standard video signal to reference to. The typical sFMs transponder has *no video signal present;* there is no reference.

If chasing a TV signal across the dial was no fun, chasing an audio signal across the dial is no-fun-squared. The TV signal is very "wide;" the audio signal is very "narrow." It is far easier to keep your eye on a moving basketball than a moving ping-pong ball. In simple terms, without AFC the ARO receiver constantly moves about and the desired audio signal is bouncing all over the dial. *Not a saleable product.* See **Figure 4.6.**

Clearly, an ARO AFC was required, but how to do it without driving the costs up very high? One technique is to build your own local reference signal into the receiver. The internal reference signal approach works on the theory that it is not the satellite signal that is drifting or moving about; it is the receiver itself. To be specific, the "local oscillator" is not totally stable with changes in operating temperature, nor time. So if you cannot "borrow" the stability of the incoming video carrier signal as a reference, you build your own reference.

The next problem is the "dither" or energy dispersal waveform. This is a form of "scrambling" placed on "all" carriers (says the FCC) to make terrestrial microwave and satellite microwave compatible. They both share the same band, as TVRO installers who have had "TI" (terrestrial interference) problems are aware. The 30 Hz energy dispersal waveform drives the relatively narrow (in bandwidth) audio carriers "crazy." It complicates creating a good, stable, "AFC" system. It must be eliminated (i.e., clamped) in the receiver or nothing will work properly.

With this we can move on to other equipment areas in SCPC and look at some of the available SCPC receivers being sold today. This equipment will handle the narrow band transmissions. See **Figure 4.7** and **Figure 4.8.**

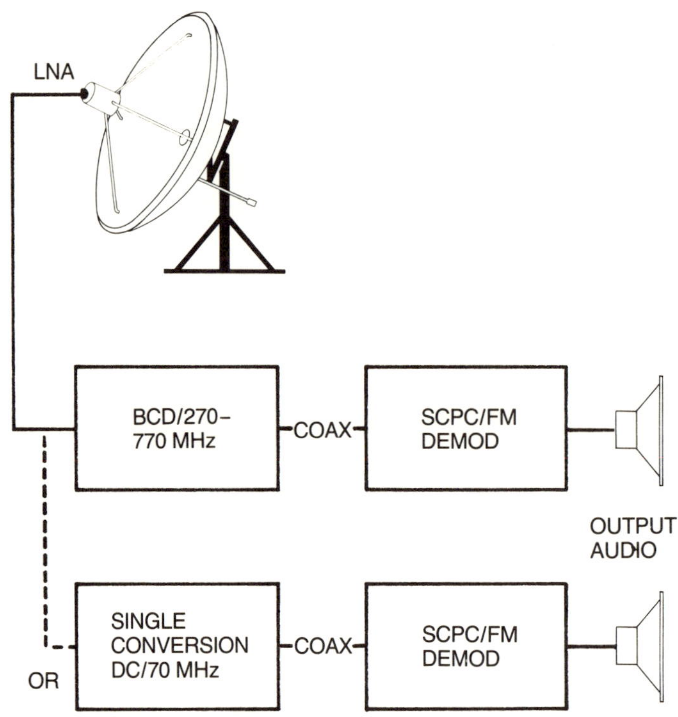

Figure 4.4 *Basic FM/SCPC system operates as an audio-only TVRO.*

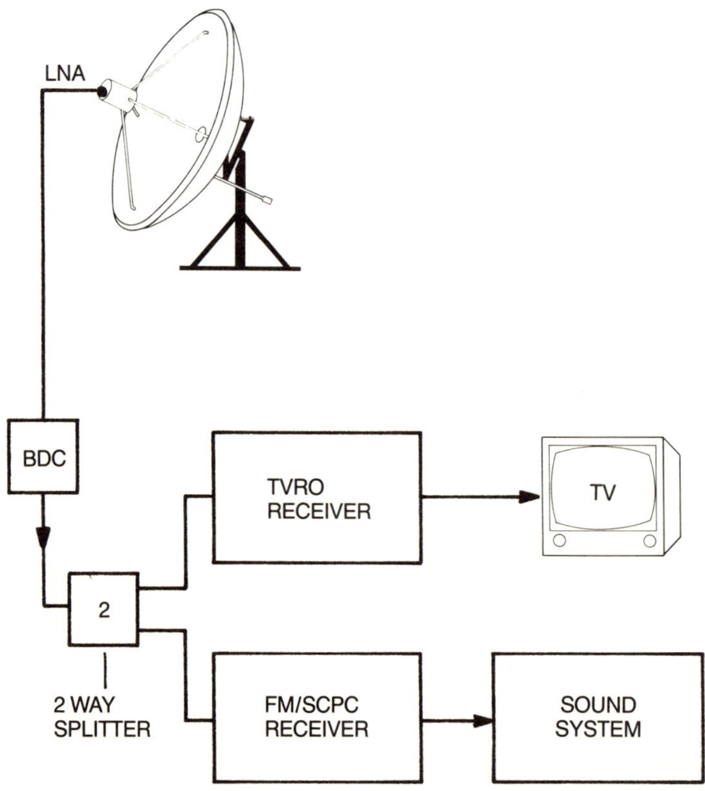

Figure 4.5 *With appropriate block down conversion receivers (AVCOM 66 series) twin receivers can be driven with single block down converter.*

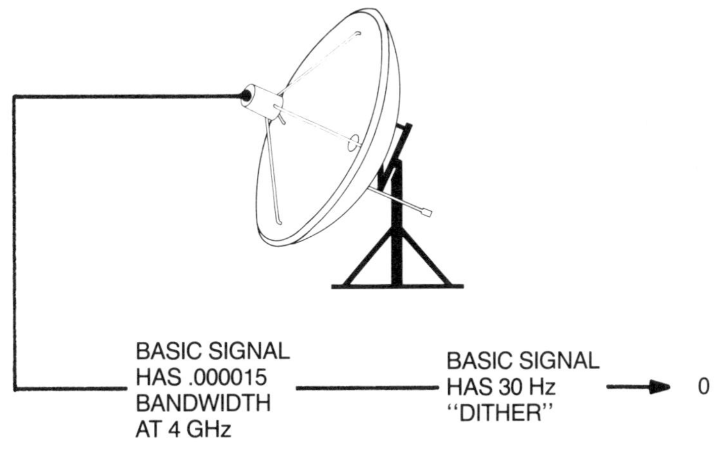

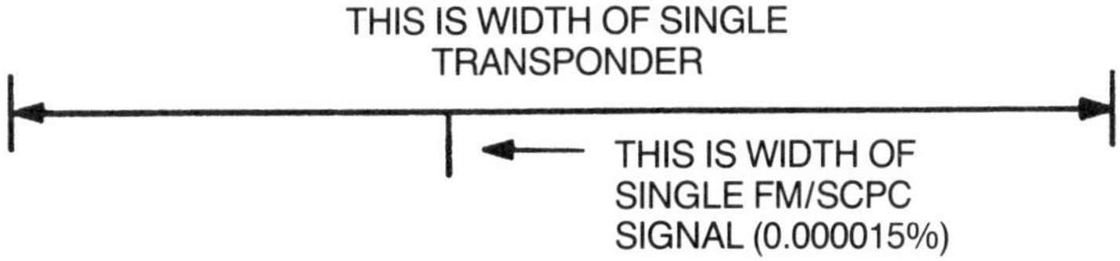

Figure 4.6 *The stability problem.*

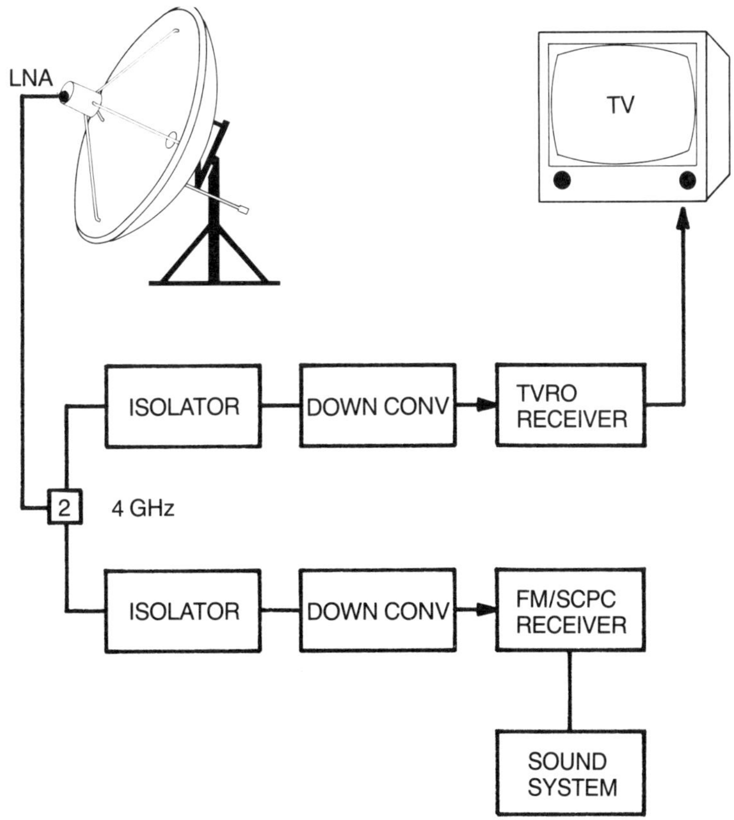

Figure 4.7 *Using single conversion receivers, dual down converters and isolators are required.*

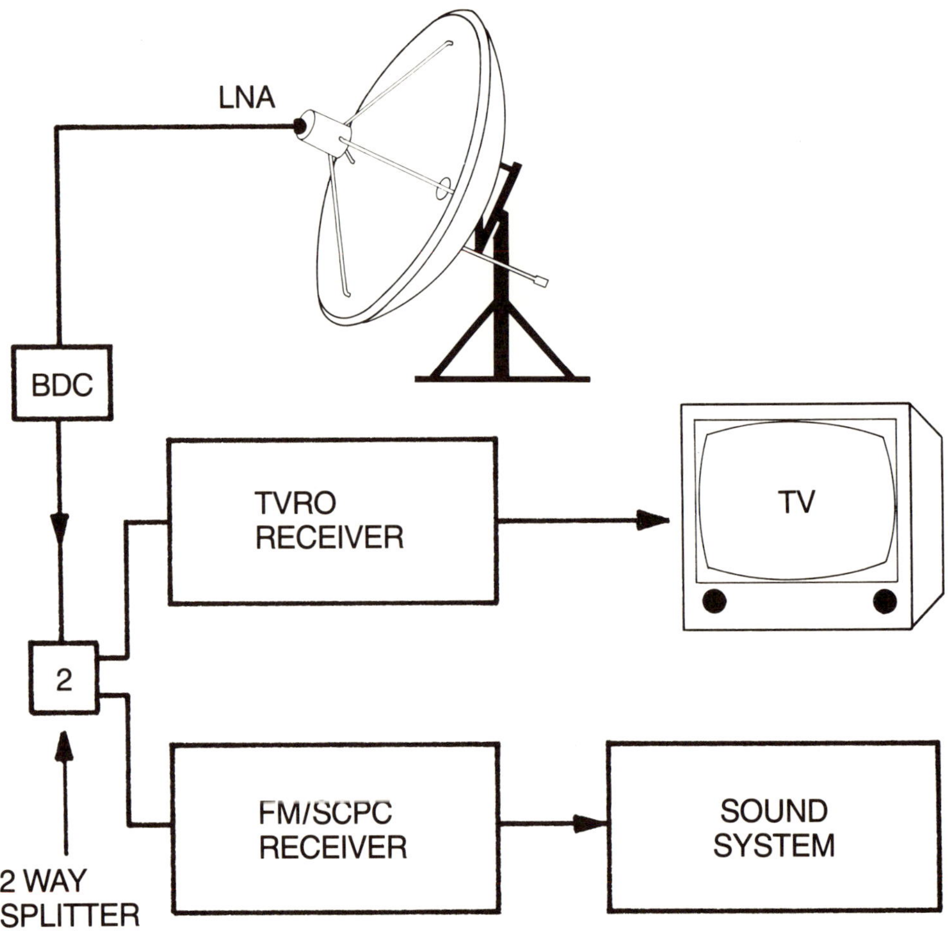

Figure 4.8 *With appropriate block down conversion receivers (AVCOM 66 series) twin receivers can be driven with single block down converter.*

SCPC/FM EQUIPMENT — GENERAL

Most SCPC equipment has been manufactured for the commercial market by several firms in the satellite data and audio field. These firms manufactured for a very narrow market, for a highly specialized use, (actually a single purpose end use) therefore the equipment was expensive and often difficult to obtain from stock. In the last few years, several manufacturers of audio and data SCPC equipment have set up dealers for their equipment, and it is not uncommon to be able to purchase all types of electronics off the shelf or within several days with notice.

The data field now offers new opportunities for the dealers who have experience in erecting and installing this grade of equipment for the end user.

Some of the leaders in this field are:

- Harris Corporation, Melborne, Florida
- Comtech Data Corporation, Scottsdale, Arizona
- Scientific Atlanta Corporation, Atlanta, Georgia
- Modulation Associates, Mountain View, California
- AVCOM of Virginia, Inc., Richmond, Virginia

HARRIS CORPORATION

Harris Corporation manufactures a wide line of satellite equipment for receive-only stations for news-wire teleprinter, news audio, stereo or monaural program audio, photo-facsimile and digital data.

To an ever-growing extent, news-wire services, radio network programming and other special communications are being distributed nationally by satellite. Newspapers, radio stations, television newsrooms and stockbroker's offices increasingly receive information and programs via satellite receive only earth stations.

Harris has developed the series 6015 satellite receive-only terminal to meet the needs of these users of satellite services. This series is field proven; over 1500 terminals have been installed for major news service, state and regional radio networks across the nation. This terminal includes a 3 meter (10 foot) diameter receiving antenna, mount, all electronics, and the interconnecting cable. See **Figure 4.9.**

HARRIS SCPC/FM NETWORKS

For light traffic (primarily telephony) from many stations, Harris Single Channel Per Carrier FM satellite networks provide optimum service. Harris SCPC/FM equipment is being used for national telephony networks in Nigeria, Sudan and Saudi Arabia. An SCPC/FM network is also being built by Harris for the rural telephone system of Argentina. A similar network is planned for communications during construction and operation of the Northwest Alaskan Gas Pipeline.

In a SCPC system, each telephone signal has its own independent channel unit with separate transmit and receive carriers. These carriers are present only when a voice (or wireline modem) signal is being transmitted, and drop out during pauses. This allows the limited transmit power of the satellite to be used only for active carriers, increasing the effective signal strength.

Harris builds a Demand Assigned Multiple Access (DAMA) control system for SCPC/FM networks. This system places all channel unit frequencies under control of a central computer. Channel unit frequencies are programmed, as required, to set up calls between any two channel units in the network. DAMA control increases the traffic capacity of the network and reduces the number of channel units needed at each station.

For special requirements, Harris TDM and SCPC/FM equipment can be combined in a single network. The TDM equipment can be used for heavier traffic links, for digital data signals, or for requirements where voice or data must be encrypted. See **Figure 4.10.**

RECEIVE-ONLY NETWORKS

Satellites can provide economical distribution of programs or information from a single source to many receivers scattered over an extensive territory. Television programs, newswire services, stock and financial data, radio network programs and similar information programs are being distributed nationwide by U.S. satellites; for example, Harris builds an extensive line of equipment for these receive-only networks, including uplinks and receiving stations for television, program audio and newswire. The satellite terminals that United Press International subscribers use to receive their audio and news teletype are built by Harris. These terminals are low in price, and are easily adaptable to distribution of music, computer data or other types of information. A 10-foot antenna connected to a small tabletop receiver comprises a complete terminal.

For television, Harris builds complete systems, including antennas, low noise amplifiers, and video receiving equipment.

Figure 4.9 *Harris satellite Receive Only terminal, series 610 S.*
Courtesy Harris Corp.

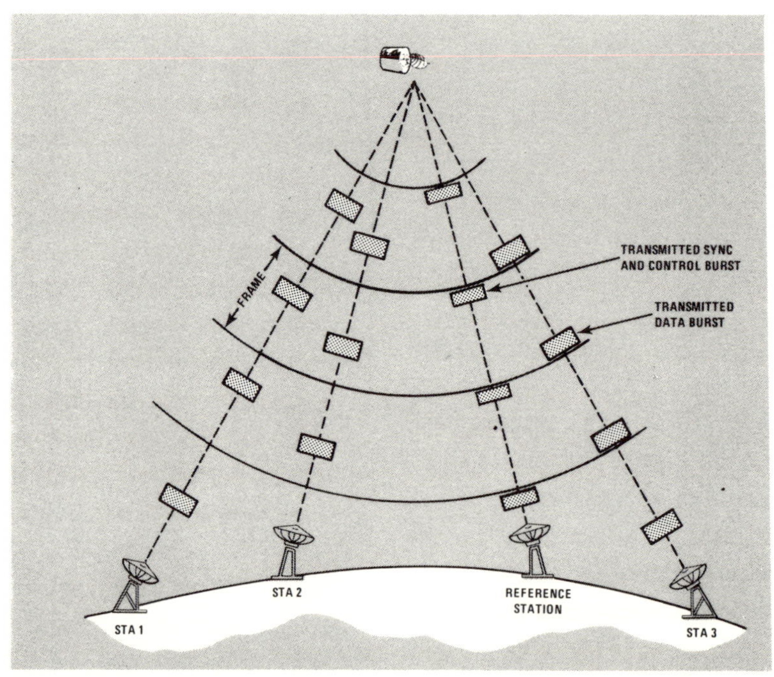

Figure 4.10 *Earth station transmit signals on DAMA SCPC/FM network.*
Courtesy Harris Corp.

TDMA IN BRIEF:

TDMA is the most advanced networking method being used to communicate over satellites. TDMA is a digital transmission technique. All signals — voice for telephone calls, computer data, teletype, facsimile and television — are converted to digital bit streams. These bits are combined into a single stream for transmission, then separated at the receive station into separate bit streams, and decoded to obtain the original message signals.

The most significant difference between TDMA and other techniques, however, is not its digital format, but the rather elegant method used to transmit the signals. In a TDMA network, each station takes its turn in transmitting information through the satellite to all other stations, in "round robin" fashion. Each station sends this information in short bursts at extremely high rates — up to sixty million bits per second. The "round robin" switch from station to station occurs rapidly, so each station transmits and receives signals many times each second. Because the switching is so rapid, the two-way flow of signals seems to be continuous to those using the network. A telephone call over a TDMA network, for example, sounds no different than any other telephone call.

The major advantage of TDMA is its flexibility. Changing the amount of traffic between stations requires no channel card changes. Instead, software-controlled division of time slots is changed to match the new traffic requirements. If required, this allocation of time slots can be done dynamically by the network master control computer, to match time-of-day changes in traffic. Even if additional stations are added to the network, only simple software changes are required at existing sites.

If full transponder traffic (60 Mb/s) is being transmitted by the network, another advantage is the extra output power available from the satellite. Because only one burst at a time is transmitted by the satellite, the transponder can be driven near saturation without creating distortion and interference between signals.

HARRIS TDM NETWORKS

Harris TDM networks for voice, data and video teleconferencing use digital transmission over multiple carriers. These networks are designed to be cost effective for light to medium traffic. Continuous carriers are used for transmitting and receiving data at each earth station. Data rates up to T1 (1.544 megabits per second) per carrier are standard; higher rates may also be used for special applications.

Cost analysis of Harris TDM networks by independent auditors have shown considerable cost reduction compared to standard common carrier services. The networks are designed to provide user organizations with increased control over both traffic and cost for data and telephone signals. See **Figure 4.11**

NEWS SERVICE PRODUCTS

Increasingly, newswire services, radio programming and other special communications are being distributed nationally by satellite. Newspaper offices, television newsrooms, radio stations, and stockbrokers' offices are receiving news, programs and internal information on their own satellite receive-only earth stations. In response to the growing demand for multipoint distribution of voice and data via satellite, Harris Satellite Communications Division has developed a receive-only terminal tailored to meet specific news service customer requirements.

The Harris voice/data terminal, called the 6100 Series, comprises a 3-meter Delta Gain receive-only antenna, a voice/data receiver and related electronics. The voice/data receiver is configured with a signal down converter and appropriate plug-in modules to receive newswire, program audio, data and facsimile, in virtually any combination. The system provides superior signal quality and reliability for radio news, teletype, digital information, stereo audio and a growing range of other services, while eliminating the need for leased telephone landlines.

The Associated Press and United Press International distribute newswire services to subscribers via Series 6100 terminals. In addition, a majority of state radio networks in the United States have chosen the Series 6100 for programming distribution. Over 1500 U.S. locations are currently receiving news, sports, and information on the Series 6100 system. See **Figure 4.12.**

RECEIVER:

The Model 6550 Receiver is designed for tabletop use or rack mounting. It is about the size of an FM tuner. Model 6550 operates from conventional 115 Vac, 60 Hz power (250 watts maximum). The receiver

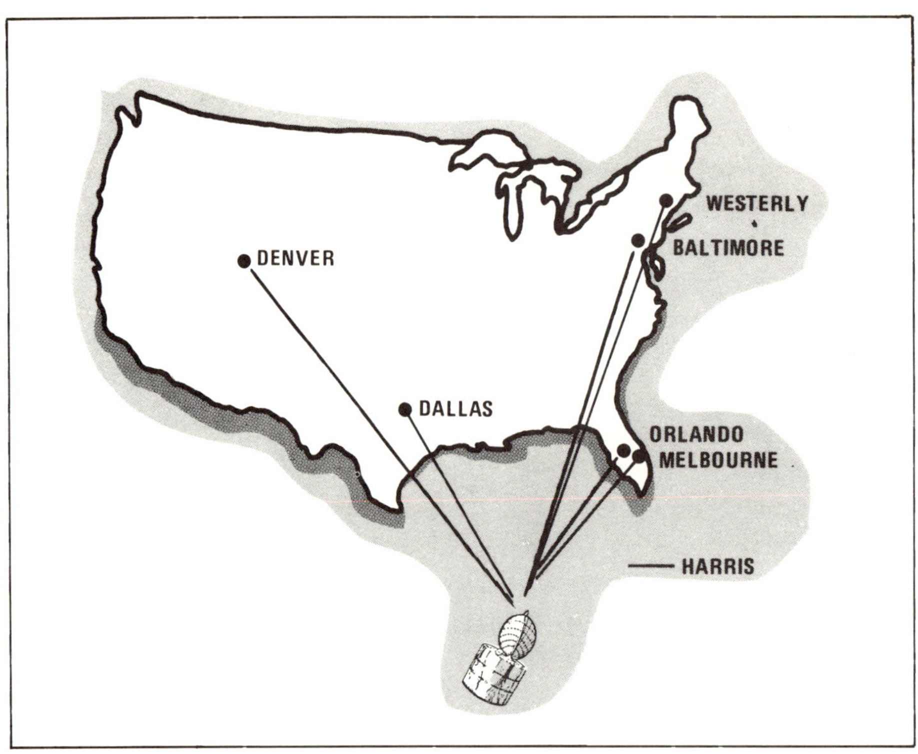

Figure 4.11 *Harris TDMA network.*
Courtesy Harris Corp.

Figure 4.12 *News services distributed via satellite.*
Courtesy Harris Corp.

contains a power supply module, a down converter module and six slots for channel processor cards. Output connections from the receiver to associated equipment are made by the twisted pair wire, connected to the terminal blocks on the receiver's rear panel.

CHANNEL CARDS:

Several configurations are available, depending on network requirements. In many cases, these cards are supplied by the news service or broadcast network to their authorized subscribers. The Model 6550 Receiver is compatible with cards supplied by major U.S. news services and radio networks.

In most U.S. locations, by adding a signal splitter and satellite video receiver, the same earth station can be used to receive TV signals from the same satellite as audio or newswire services. See **Figure 4.13.**

ANTENNA

The Model 5115 Antenna may be mounted either at ground level on a reinforced concrete pad, or on a rooftop. Ground level mounting is generally recommended, because of lower cost, less microwave interference and easier installation. For rooftop mounting, a special steel load frame is installed and bolted or welded to the building framework. Because of wind loading effects on the antenna, this ring and the building structure must be capable of handling loads up to 6400 pounds at each of four attachment points.

The installation area must have a clear line-of-sight toward the satellite. Buildings, guywires, fences and trees will block signals. The entire southern sky in the arc occupied by satellites should be free from obstructions. This will allow future changeover to other satellites if required. **Figure 4.14.**

PRIVATE/CORPORATE NETWORKS

Technological advances are allowing the development of a variety of communications networks that meet specific corporate communications needs. Local area networks connect computers, telephones and information processing equipment within a business complex. Switch and access networks connect several local networks to each other. Wide area networks provide national or even global linkage of the first two types of networks. Increasingly, satellite communications networks are the cost-effective solution to wide area network requirements.

Private satellite networks provide optimum performance for corporations with multiple plant sites, varying traffic patterns, and an assortment of information requirements, such as voice, video, data and facsimile. Common and independent carriers that address the corporate communications market also benefit from the improved performance and reduced costs provided by satellite networks. Harris Satellite Communications Division produces private satellite networks that meet all of these communications needs. Harris Corporation's own private satellite network, for example, provides voice and data communications between three U.S. locations. A Harris-built switching system provides automatic least-cost routing of all long distance telephone calls for twenty Harris divisions in the vicinity of the main sites. See **Figure 4.15.**

TELEVISION BROADCAST NETWORKS

Television network programming is increasingly being distributed by satellite in the United States and worldwide. In the U.S., the National Broadcasting Company (NBC) is converting to satellite communications for distribution of all television network programming to NBC affiliates. Harris Satellite Communications Division was awarded the contract to design and install the ground segment equipment for this network.

The new distribution network will operate at Ku-band to provide reduced terrestrial interference, and to allow location of earth stations at urban sites. The equipment designed by Harris features transportable antennas for on-location broadcast of news and sports events, fixed earth stations for transmit as well as receive capabilities, and extensive software design for a satellite network control and monitor system.

Earth stations at the NBC affiliates will be fully automated to allow unattended site operation. The stations will be controlled from a master satellite network management system to provide the most cost-effective and accurate operation. The satellite network will also provide high speed data communications between affiliates and NBC headquarters in New York City. See **Figure 4.16.**

Figure 4.13 *Harris 6550 receiver with channel cards.*
Courtesy Harris Corp.

Figure 4.14 *3 meter delta gain antenna.*
Courtesy Harris Corp.

Figure 4.15 *Satellite network control center.*
Courtesy Harris Corp.

Figure 4.16 *Television network Receive Only configuration.*
Courtesy Harris Corp.

DIGITAL EQUIPMENT

Harris products for digital communications include digital BPSK and QPSK modems for data rates from 1200 bits per second to 60 megabits per second, forward error correcting encoder-decoders that significantly improve modem performance, voice digitizing and encrypting equipment, and high speed multiplexer/demultiplexers. Harris Satellite Communciations Division digital products are in wide use throughout the world in corporate networks and domestic satellite systems. See **Figure 4.17.**

CONTROL AND MONITOR EQUIPMENT

As satellite communications networks become more complex, the need for sophisticated control systems increases. Harris Satellite Communications Division builds a Communications Management System (CMS) which provides master station monitoring of all network equipment, remote switching and recording of network operations. With CMS, remote stations operate unattended, under control of the master station. The new NBC satellite network uses a specialized Satellite Network Management System. (SNMS). See **Figure 4.18.**

SCPC

For light voice or data traffic (primarily telephony) between many stations, Harris Single Channel Per Carrier or SCPC satellite networks provide optimum service. Harris SCPC/FM equipment is being used for national telephony networks in Argentina, Nigeria, Sudan, and Saudi Arabia. The SCPC system is simple and cost-effective because each voice or data circuit operates independently. Harris builds a complete line of this flexible, low-cost SCPC equipment, including a Demand Assigned Multiple Access (DAMA) control system, which places all SCPC channel frequencies under the control of a single computer. DAMA control increases the capacity of each network and reduces the number of channels required at each station. See **Figure 4.19** and **Figure 4.20.**

MICRODYNE CORP. SINGLE CHANNEL PER CARRIER SYSTEM

Microdyne's 12 Foot Antenna and Single Channel Per Carrier (SCPC) Terminal meets industry demand. This advanced system delivers higher systems margins to meet the increased activity and demand of the industry's analog and digital subscribers.

Microdyne's 12 foot antenna features a minimum gain of 41 dB at 4 GHz with easy polarization adjustment. The 12 foot antenna is a three-piece molded fiberglass unit with a precision reflector surface. Mounted on a galvanized steel pedestal, the antenna has a full 360 degree azimuth range, with a 10 degree to 65 degree adjustable elevation. Microdyne's 12 foot antenna is designed to withstand winds up to 125 mph. It weighs 500 pounds.

The down converter electronics are provided in two configurations. The SCPC system is supplied as an antenna mounted down converter (as shown), or as an individual component system for indoor, rack mounted equipment.

Features: 3.7 to 4.2 GHz input—70 MHz output—antenna mounted or rack mounted down converter — crystal controlled conversion — 55 to 85 MHz output bandwidth — prime focus feed. See **Figure 4.21.**

This SCPC system gives the choice of antenna mounted down converter (DC4–70) which is a rugged fixed unit, **Figure 4.22,** or the 1100 FFC–X1 (RDC) L rack mounted SCPC down converter for multiple transponder reception. See **Figure 4.23.**

1100–PCDR SCPC DEMODULATOR (FULLY TUNABLE)

Microdyne's 1100 PCDR Single Channel Per Carrier (SCPC) demodulator provides the radio broadcaster with access to a wide range of audio programming services available via satellite. The PCDR demodulates and processes the 70 MHz IF analog signal supplied by a companion 1100 FFC–X1 (RDC) L SCPC Down Converter.

Microdyne's standard 1100 PCDR (3) Demodulator provides front panel switch selection of Narrow Bandwidth (0 de-emphasis), Wide Bandwidth (25 μs de-emphasis), and Wide Bandwidth (no de-emphasis). The narrow bandwidth mode is used to process transmissions from low power transponders (typical 10 dBw footprint). The PCDR is designed for applications where multiple SCPC networks must be economically

Figure 4.17 *Digital BPSK and QPSK modems for data rates from 1200 bits per second to 60 megabits per second.*
Courtesy Harris Corp.

Figure 4.18 *Satellite network control and management system.*
Courtesy Harris Corp.

Figure 4.19 *For light data or voice service between many stations, SCPC networks provide optimum service.*
Courtesy Harris Corp.

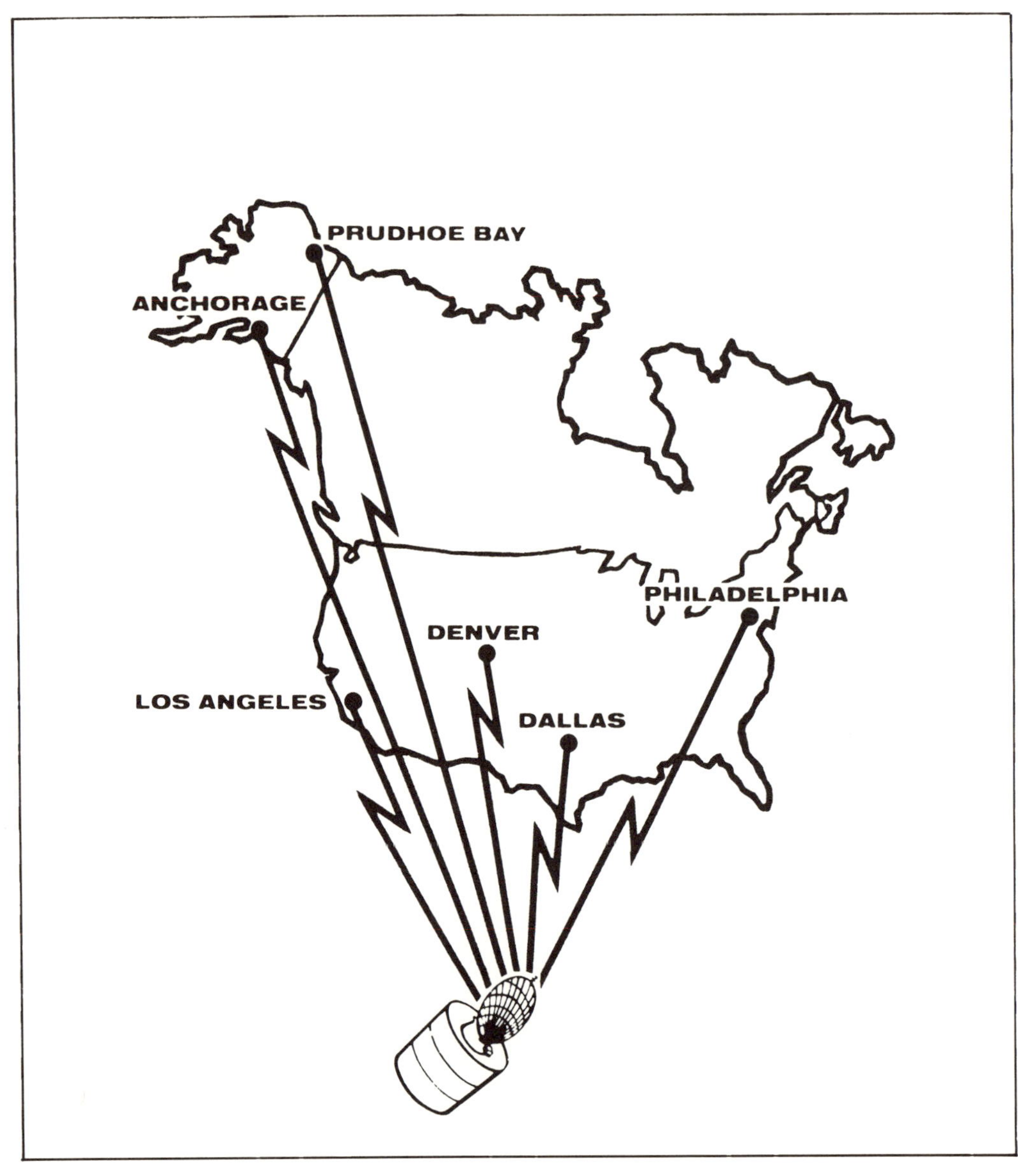

Figure 4.20 *The SCPC system is simple and cost effective because each circuit operates independently.*

Courtesy Harris Corp.

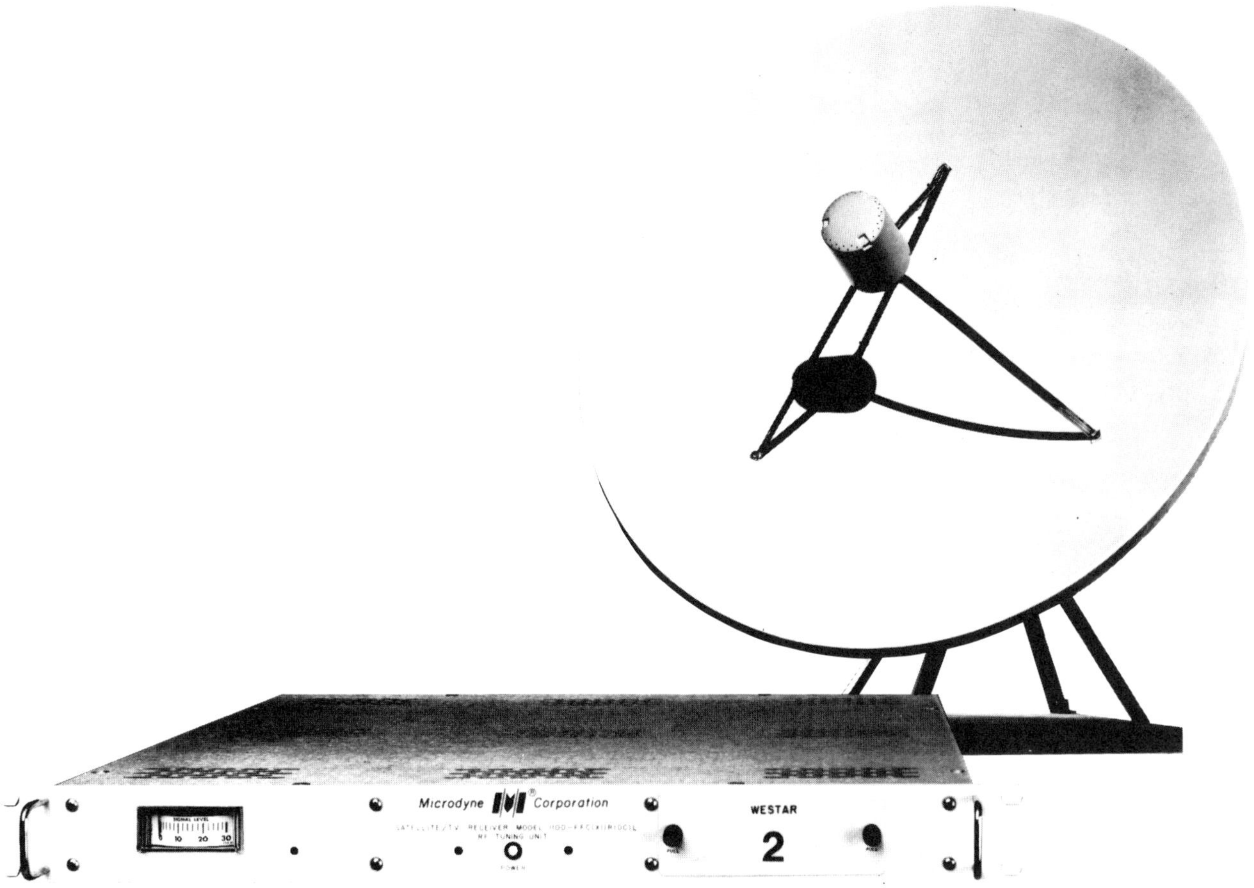

Figure 4.21 *Microdyne single channel per carrier system with 12 foot antenna with rack mounted down converter.*
Courtesy Microdyne Corp.

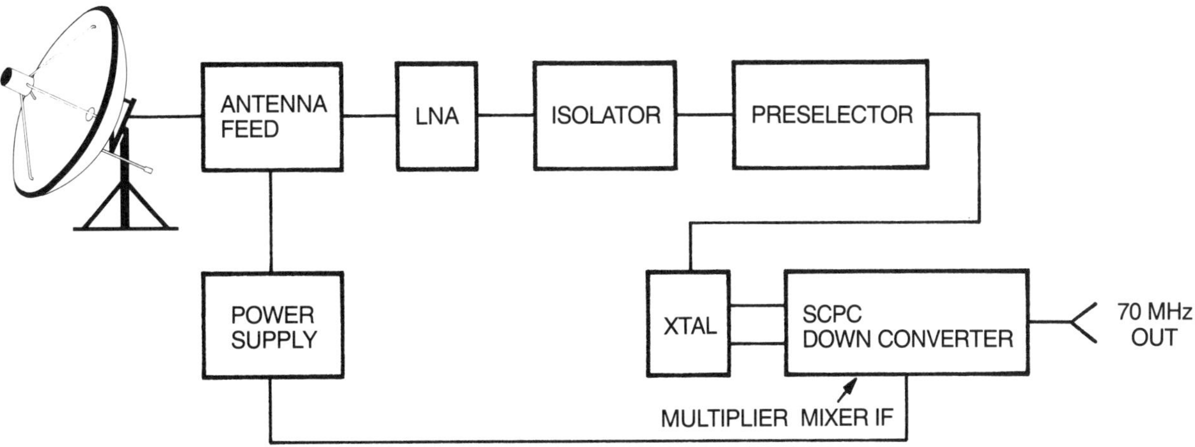

Figure 4.22 *DC4–70 rugged antenna mounted SCPC down converter.*

accessed. Narrow bandwidth satellite audio transmissions are very cost effective compared to utilizing phone lines for servicing large numbers of radio stations.

The standard 1100 PCDR single bandwidth demodulator (wide or narrow bandwidth) can be tailored to meet specific requirements including pre and post detection bandwidths, de-emphasis and expanding ratios.

Front panel step tuning provides fast and precise selection of SCPC radio channels. The four/position frequency thumbwheel switch is used to select the desired frequency while observing the tuning indicator (AFC loop stress meter), which indicates when the input carrier frequency is centered within the AFC tracking range. When a carrier is not present, the audio output is muted. An input directional coupler provides a rear panel loop through RF output without using splitters or power dividers.

Featuring a standard 1:3 expansion ratio and a selectivity of 40/190 kHz, input levels as low as −65 dBm provide a +18 dBm output. Other features include: an audio gain adjust, headphone monitor jack, carrier indicator, tuning indicator and rear panel contact closure output for remote loss of carrier alarm. One 600 ohm balanced audio output is standard with up to three additional buffered and fully isolated outputs optionally available. **Figure 4.24.**

Other features are: synthesized 10 kHz step tuning—50 Hz to 7.5 kHz and 15 kHz audio ranges—600 ohm balanced audio outputs — audio muting — carrier tuning indicators.

COMTECH DATA CORP. SERIES 300 RECEIVE-ONLY SCPC FOR NEWSWIRE, AUDIO AND DATA RECEPTION

The Series 300 Receive-Only Terminal has been designed to meet the needs of the user of satellite services. It allows the reception of Single Carrier per Channel (SCPC) audio, Multiple Carrier per Channel (MCPC) audio, digital data, composite video baseband, and subcarrier audio. Up to 6 crystal controlled transponder segments may be selected.

The system G/T performance is rated at 21 dB and allows a low satellite EIRP of 10 dBw for a 7.5 kHz channel. As a result, only 1% of a satellite transponder's bandwidth is required to achieve the same performance provided by other systems requiring up to 10%.

Four components are used in the standard configuration. These consist of a 3.8 meter figerglass antenna, a 100 degree K low noise amplifier (LNA), an antenna-mounted down converter, and a Uni-Shelf electronic package. Plug-in modules allow the system to be configured for a mixture of received signal types (SCPC, MCPC, data, etc.). An optional up-link electronics package allows the system to provide both up-link and down-link capabilities. **Figure 4.25.** Complete system packages are available and include the antenna, mount, and all electronics. Applications for this system: Mono and stereo audio, news audio, SCPC/MCPC subcarrier audio, teleprinter wire-services, photo-facsimile and digital data. See **Figure 4.26.**

The Control Unit features the COMTECH Uni-Shelf concept. This feature allows the user to configure a system that matches his unique requirements. It is designed to mount in a standard 19″ rack, and requires only 7″ of front panel space. See **Figure 4.27.**

Several types of channel outputs are available. These include audio SCPC, audio MCPC, digital data, composite video, and audio subcarrier channels.

Audio SCPC: Both voice grade and wide bandwidth capabilities are available. The latter provided essentially flat response from 50 to 15 kHz, with less than a .5% distortion.

MCPC: Program audio up to 15 kHz and Cue or Data multiplexed in a 4 kHz band above.

Digital Data: Differentially encoded, QPSK-modulated carriers may be received. Data rates up to 1.5 Mbps are supported and a CCITT V.35 interface is provided.

Composite Video: A baseband composite video output is provided for use with Series 300 subcarrier demodulators or external equipment.

Audio Subcarrier: Demodulation of video subcarrier pairs from 5 to 9 MHz is provided. The outputs are essentially flat from 50 to 15 kHz with less than .5% distortion. Companding and stereo matrix decoding is available.

RCV 360 SATELLITE RECEIVER

The RCV 360 Satellite Receiver allows the reception of SCPC, MCPC, and digital program material. Up to six satellite transponder channels may be selected using either a local or remotely-located switch.

The down converter is an environmentally sealed unit which mounts at the antenna. It converts the 3.7

to 4.2 GHz output of the LNA (not included) to a 70 MHz IF signal for use by the Uni-Shelf demodulator. The 70 MHz IF is transmitted to the demodulator over low-cost coaxial cable. This same cable is used to provide low-voltage DC power to the down converter.

The Uni-Shelf demodulator is designed for mounting in a 19″ rack. It is 7″ high and consists of a card shelf, plug-in power supply module, control board, and from one to six demodulator boards. A protective hinged, front panel allows easy access when open.

The SCPC demodulator allows a choice of three audio bandwidths and two companding ratios. It also demodulates carrier FM deviation ratios ranging from ±15 kHz to ±75 kHz.

The MCPC demodulator allows operation with a program audio channel of up to 15 kHz and a multiplexed cue or data channel occupying an additional 4 kHz of baseband spectrum.

The data demodulator uses an industry standard CCITT V.35 data interface. It provides demodulation of digital data at rates of up to 1.5 Mbps with a bit error rate of 10^{-7}.

The number and types of demodulators used will depend upon the particular application. However, additional channel capacity may be added simply by inserting additional demodulator boards and making rear panel connections. A 6-channel capacity may be provided by a single RCV 360 receiver. See **Figure 4.28.**

AVCOM, INC.
AVCOM SCPC-100 PROCESSOR (RECEIVER)

AVCOM's SCPC-100 Intermediate Frequency Processor is used with one or more RDC-100 high stability down converters to convert single channel per carrier (SCPC) satellite transmissions to the 88-108 MHz band. One or more FM receivers are used to select any desired SCPC signal(s) from the down converted 20 MHz wide frequency block. See **Figure 4.29.**

SCPC transmissions are similar to those from FM broadcast stations except for a much higher frequency (3.7-4.2 GHz band) and they come from satellites in space with very weak signal levels. A 6 foot diameter or larger dish antenna and 120 degree K low noise amplifier are usually sufficient to receive the Westar 3 and Westar 4 transmissions on transponder 3 (2D).

The RDC-100 down converter is connected to the LNA via a length of coax. Location of the RDC-100 is important to minimize frequency drift. Even though the RDC-100 down converter uses a high stability oscillator, temperature changes will tax its ability to keep individual SCPC signals within the intermediate frequency passband of an FM tuner. This is due to the very narrow bandwidth of SCPC signals, typically 75 kilohertz (only .0025% of the 4 GHz carrier frequency)! Slight frequency shifts are compensated for by an AFC circuit in the FM tuner. Nevertheless, it is desirable to locate the RDC-100 down converter in an environmentally-protected area where there is a relatively constant temperature. A coax feedline of RG-217 will allow a cable run of 120 feet or more, RG-214 is good to about 70 feet, and ½ inch hardline will normally work to 300 feet. The cost of the coax is only a nominal amount and the improved stability, reliability, and convenience resulting from locating the down converter indoors is well worth the small additional investment. The RDC-100 has a built-in DC power block to provide power for the LNA via the feedline. Connect the terminal on the down converter marked "LNA" to the terminal marked "+18" or "+23."

The RDC-100 down converts a 20 MHz wide block of SCPC signals to the FM band, from 88-108 MHz. See **Figure 4.30** for a spectrum analyzer photo of the SCPC signals from Westar 4 transponder 3 (2D).

This IF output from the down converter is connected to the SCPC-100 IF Processor via a length of RG-59 coax. This type of coax may be run 300 feet or more. No voltage is present on the center conductor of the coax. The rear panel of the SCPC-100 has provisions for selecting power for three RDC-100 down converters. Three additional power outputs are optionally available. The front panel down converter selector, having six positions labeled A thru F, determines which terminal on the real panel becomes active with +23 VDC to power its respective down converter. Two conductor 18 gauge stranded wire is recommended to connect the RDC-100 down converters. A fuse on the rear of the SCPC-100, next to the down converter power terminals, protects the SCPC-100 power supply from accidental shorts.

A typical hookup of two RDC-100 down converters to an SCPC-100 IF Processor is shown in **Figure 4.31.** Two FM tuners are shown as would be used for simultaneously receiving two SCPC signals (as in stereo reception of left and right channels). Referring to **Figure 4.30,** the FM tuners can select any of the vertical "spikes" that represent SCPC signals by tuning to the frequency of the "spike." If there are

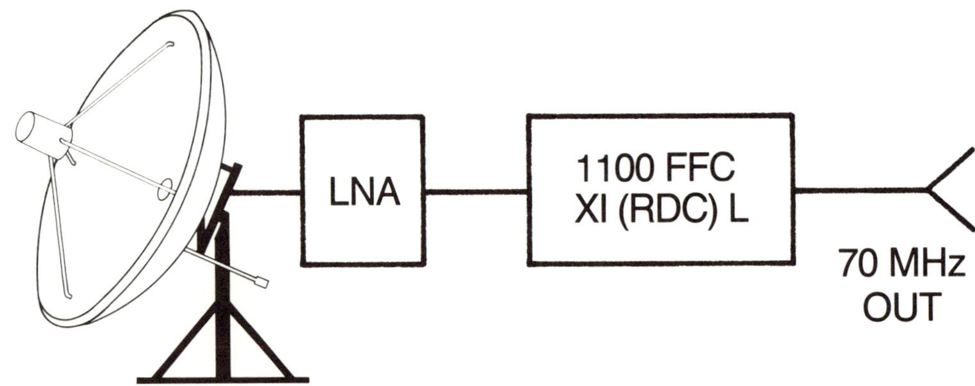

Figure 4.23 *1100FFCX-1 rack mounted SCPC down converter for multiple transponder reception.*

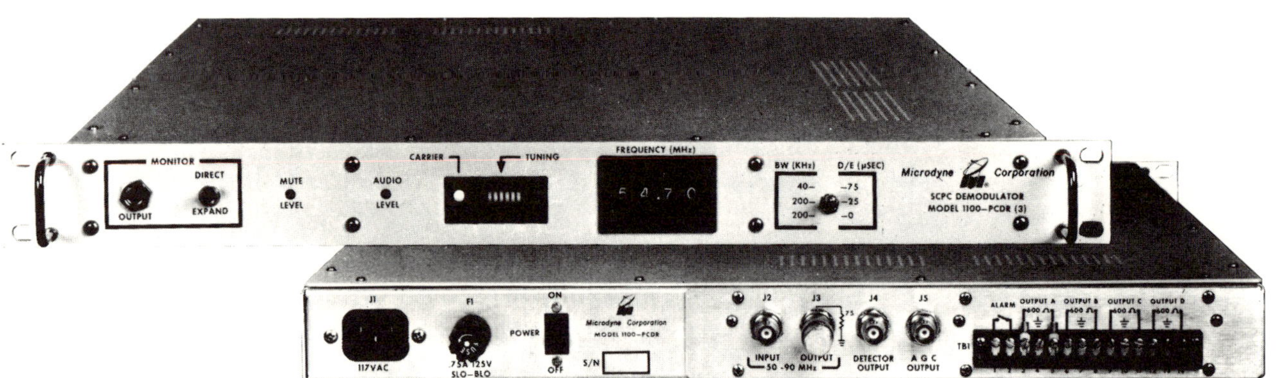

Figure 4.24 *1100-PCDR SCPC tuneable demodulator.*
Courtesy Microdyne Corp.

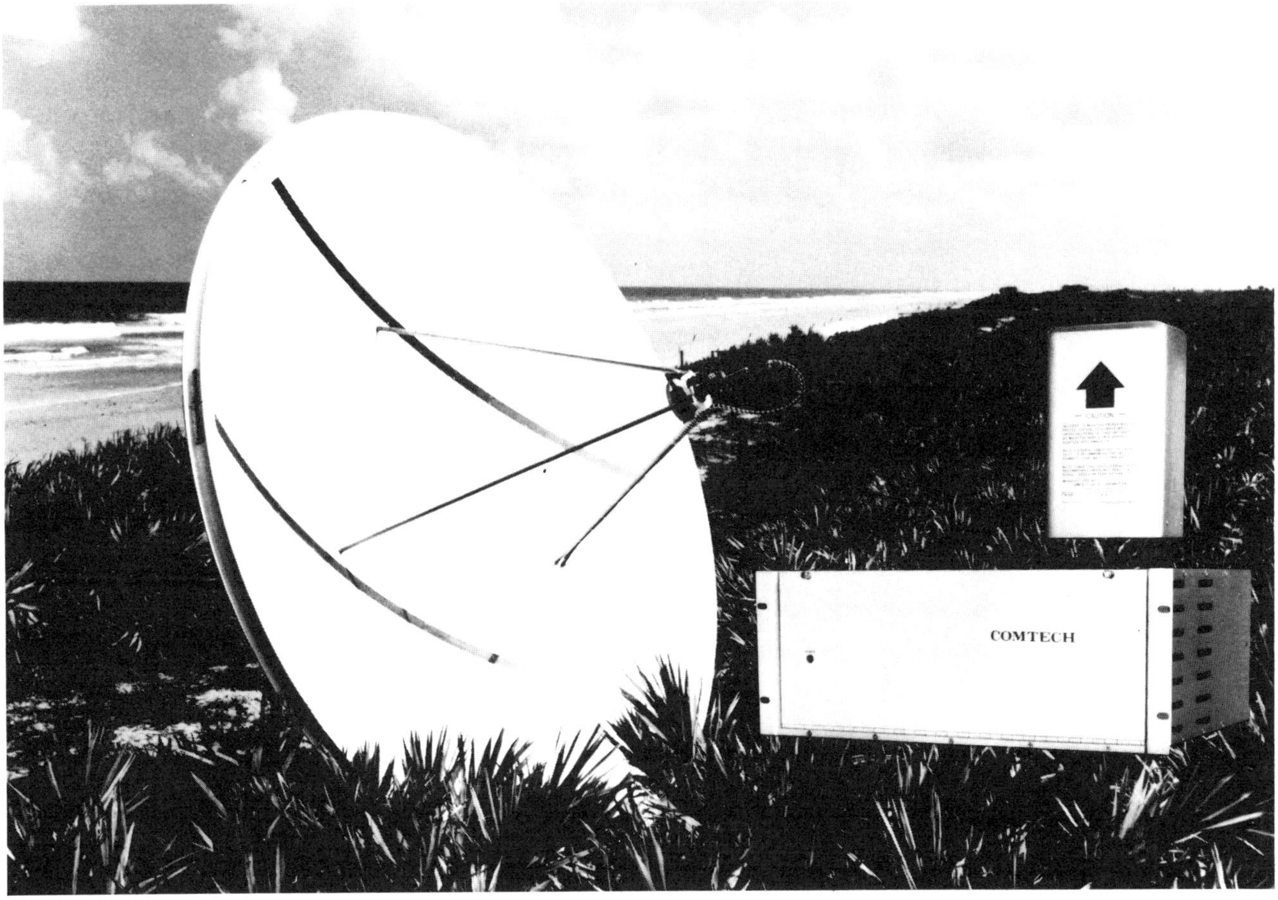

Figure 4.25 *Series 300 receive only SCPC system for newswire, audio and data reception.*

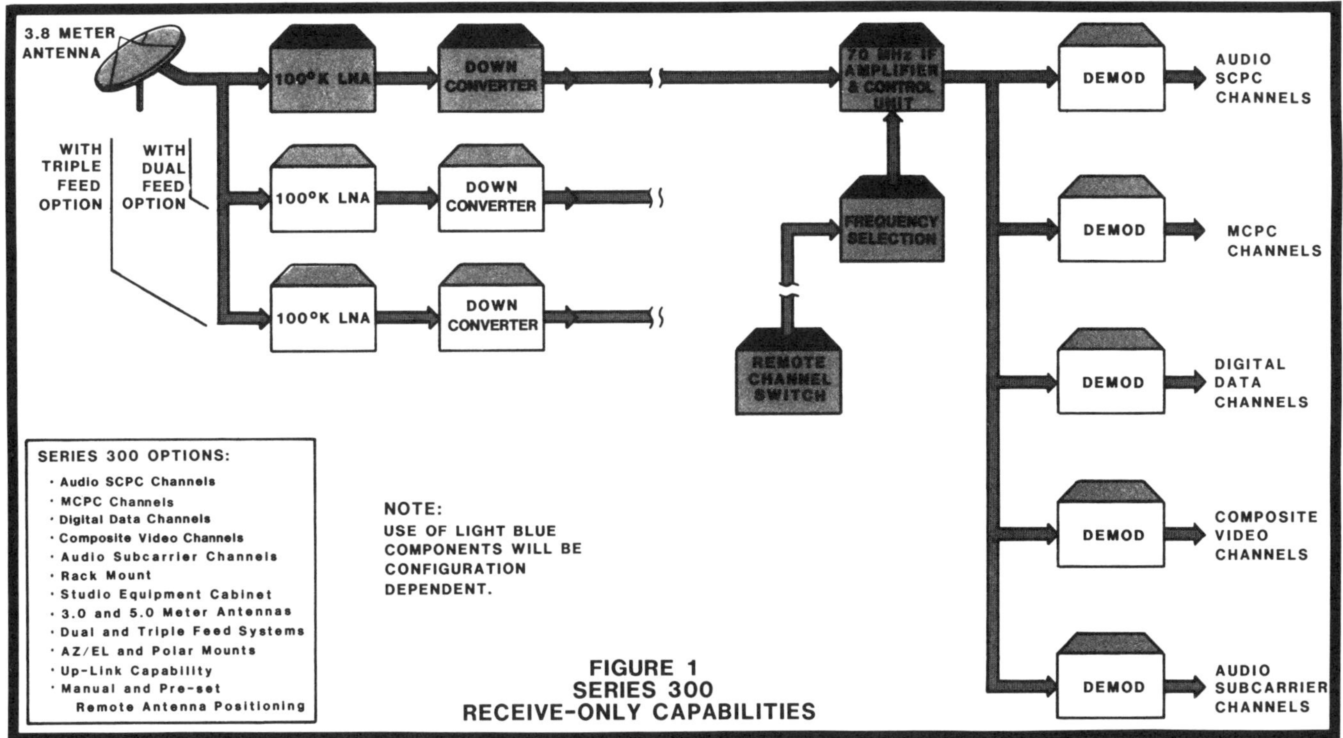

Figure 4.26 *Series 300 system—receive only.*
Courtesy Comtech Data Corp.

Figure 4.27 *Control unit features the Uni-Shelf® Concept.*
Courtesy Comtech Data Corp.

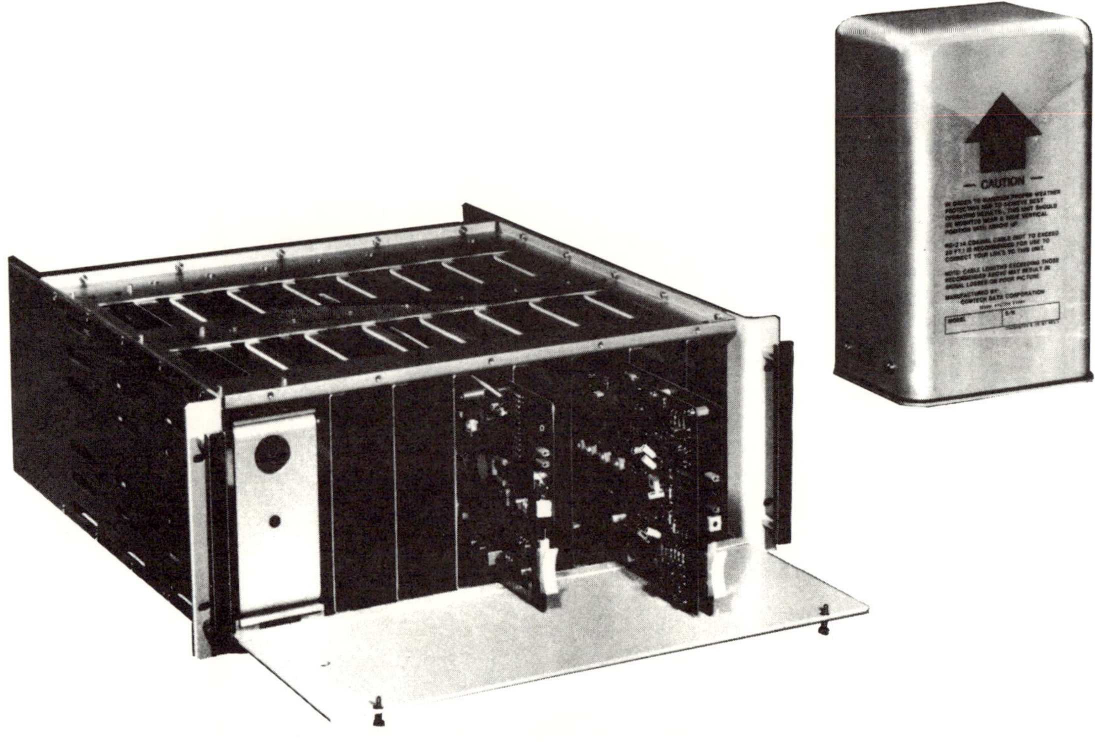

Figure 4.28 *RCV 360 satellite receiver for SCPC, MCPC, and digital data.*
Courtesy Comtech Data Corp.

10 different signals in the band, then 10 FM tuners, each tuned to a different "spike's" frequency, could be used to demodulate all of the 10 signals. Only one RDC-100 and one SCPC-100 would be needed since all 10 signals from one transponder have been block down converted to the 88-108 MHz band.

A minimum configuration would be composed of one RDC-100 down converter connected to a SCPC-100 IF Processor. No splitters would be necessary. An FM tuner connected to the IF Processor output would permit demodulating any SCPC signal, one at a time.

Figure 4.29 *AVCOM SCPC-100 unit used with RDC-100 down converter.*

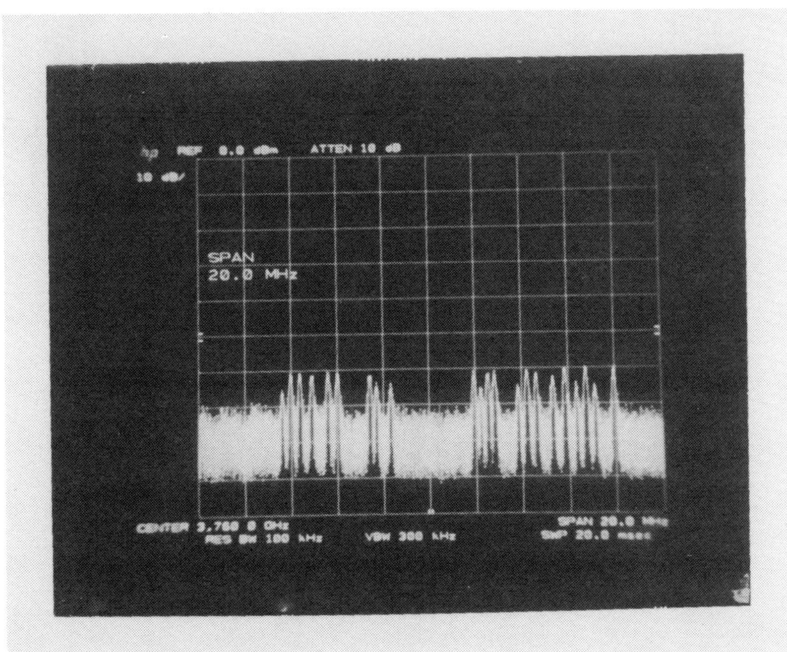

Figure 4.30 *Spectrum Analyzer photo of SCPC signals form Westar 4 transponder 3 (2D). Center frequency is 98 MHz and signals 10 MHz on either side of center are shown.*

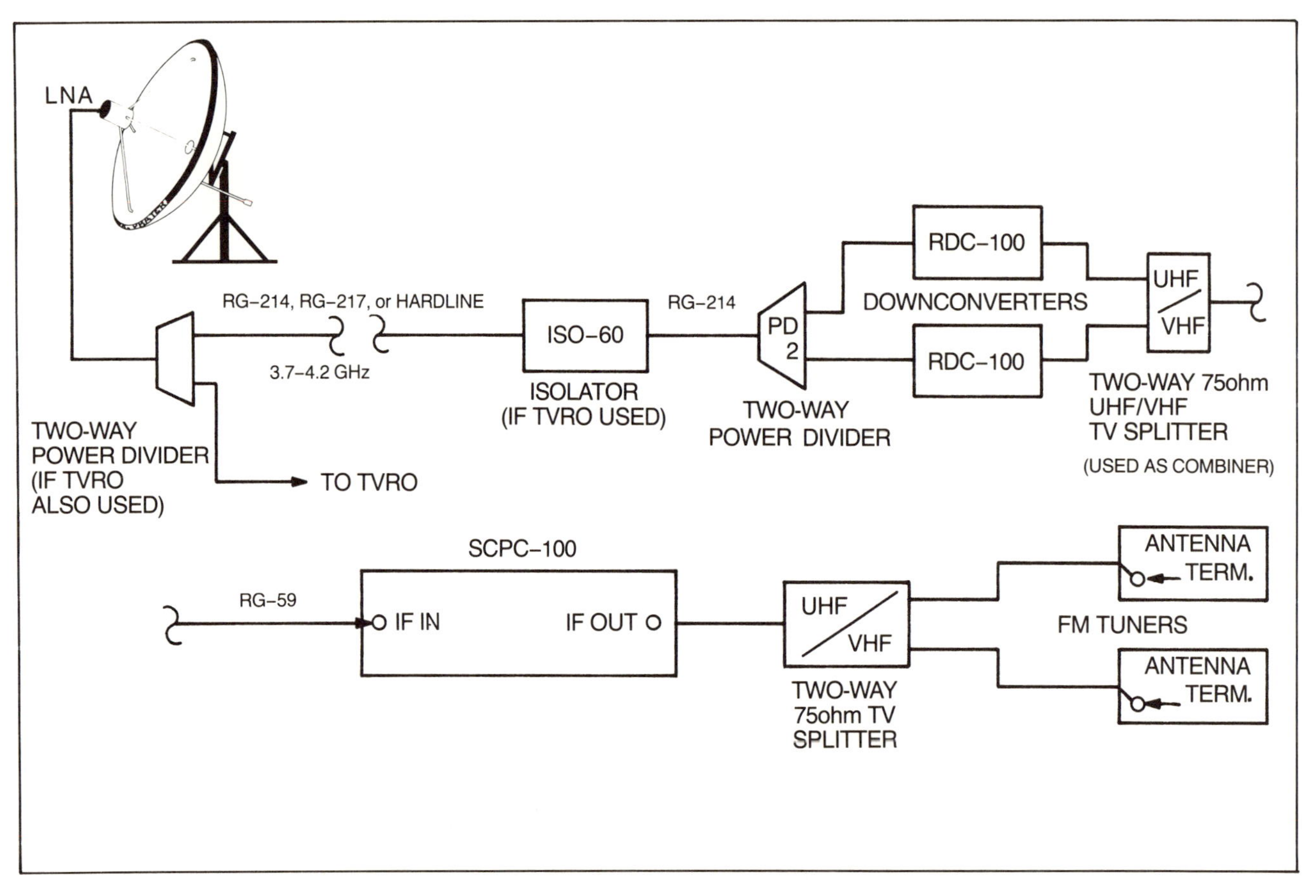

Figure 4.31 *Typical SCPC installation with two down converters, one SCPC-100 and two FM tuners.*

TUNING THE RDC-100 HIGH STABILITY OSCILLATOR

The RDC-100 down converter is normally factory tuned to a customer's specified transponder or frequency. Field tuning is accomplished by adjusting the ⅜ x 32 screw on top of the oscillator housing. Loosen the locking nut securing the ⅜ x 32 screw and adjust the screw in ⅛ revolution increments. After each increment, snug the locking nut and tune the FM tuner from 88-108 MHz. If any SCPC signals are present on the satellite, they will be heard as the adjustment procedure advances.

Lower frequency transponders will be selected with the screw protruding highest from the top of the oscillator. Higher transponders will be selected by turning the screw inward (clockwise). The tuning resolution is greater on lower transponders, and most U.S. domestic satellite SCPC signals are found on transponders 3 or 1 (24 channel numbering scheme) on Westars 3 and 4. There is a stereo radio feed (pair of SCPC signals) at the lower end of transponder 3 on Satcom 3.

SCPC-100 OPERATORS CONTROLS AND CONNECTORS

Front Panel

- Transponder Selector-switches power to any one of up to six RDC-100 down converters.
- Green LED-Indicates power is on.
- Signal Strength Meter-Indicates relative signal level.

Rear Panel

- Power Switch-Applies power to SCPC-100.
- Signal Adjust-Adjusts sensitivity of front panel meter.
- ½ Amp Fuse-Protects RDC-100 power cables.
- 1 Amp Fuse-Protects 120 VAC line circuit.
- Terminal Block, "A, B, C, Gnd"-Power source and ground return for RDC-100s.
- IF Input-Connect to RDC-100 with RG-59.
- IF Output-Connect to "F" input terminal of an FM radio with *shielded* cable (RG-59 coax).
- Line Cord-Connect to *grounded* 120 VAC outlet.

HOW TO RECEIVE INTERNATIONAL SATELLITE AUDIO ON SCPC

Most of our domestic satellites transmit the audio section of the signal on a standard subcarrier of 6.2 MHz or 6.8 MHz. This holds true for some Canadian satellites subcarriers, however, the ability to tune the continuous band of subcarriers from 5 MHz to 8.5 MHz is needed to cover all subcarriers on the domestic birds.

In the case of the Intelsat birds, many services and countries use the SCPC method of transmitting the audio section of their services; therefore SCPC equipment must be used in order to enjoy these satellites. When the SCPC method is used instead of an audio subcarrier, an equipment set-up as shown in **Figure 4.32** should be used.

The SCPC system here uses the basic AVCOM SCPC-100 unit by splitting the LNA output into the regular TVRO receiving unit and to the AVCOM SCPC-100 unit which is in turn, in line with a stable FM tuner which feeds its output to the audio input of a suitable video monitor that carries the video standard in use.

A stable FM receiver must be used with the AVCOM SCPC-100 if good results are expected. Some of the better quality FM tuners with good AFC circuits (automatic frequency control) work very well. Several suggestions for the stable FM demodulator are: (A) one Regency's new continuous coverage, digital readout scanners, such as the MX-5000 model; (B) the Sony ICF-2002 unit with digital readout and 6 memories for storage of frequencies. The new ICF-2002 also scans the frequency which will be helpful when searching for SCPC channels.

Please note that some Intelsat audio is not present on the same transponder that carries the video section, but will be located off on a different transponder on the satellite.

NARROW BAND SCPC EQUIPMENT

The narrow band transmissions we are dealing with here are truly narrow band; some are barely 4 kHz wide! That makes them tiny little fellas tucked away in a gigantic transponder, where even a 100 kHz wide subcarrier is often overlooked.

The information these contain may well be just as interesting and entertaining as that found on F3R, transponder 3. The 4 kHz transmissions are called

SSB/SCPC because they use a modulation format known as single sideband (SSB).

To tune in the SSB/SCPC signals you:

(1) Down convert your TVRO signals to standard 70 MHz, IF;

(2) Tune to the appropriate satellite and transponder sending the SCPC signals;

(3) Connect the *unfiltered video* of your TVRO receiver to the antenna input (antenna) of a quality SSB general coverage receiver—JRC-NRD 515, Kenwood 1000, Kenwood R-2000. See **Figure 4.33;**

(4) Tune the SSB general coverage receiver to O MHz up to 8 MHz to recover all of the video passband (from 0 to 8 MHz), rather than just that portion between 0 and 4.5 MHz (which normally carries the video part of the signal), since on a SSB/SCPC transponder it may be "loaded" with narrow band channels all the way to 8 MHz and above. SSB/SCPC is fine for slow speed data up to 300 baud rate. Typically SSB/SCPC data is in the range of 50, 75, 300 baud rate.

(5) From the audio out (speaker output) or on the JRC-NRD-515 audio out at 600 ohms, you connect a teletype decoder unit, M-600 RTTY demodulator at the input terminal. From the video out connector, a monitor is hooked up to give video copy of the data ("soft copy"). A parallel printer can be hooked up to the M-600 printer output to obtain hard copy. See **Figure 4.34.**

This is the basic low baud rate SSB/SCPC equipment set up. See **Figure 4.35.**

(6) After the proper equipment is hooked up, the search for data signals begins. The transponders are searched for the RTTY sounds. Some RTTY signals sound like small bells, others sound like a harsh buzzing. See **Figure 4.36.**

NARROW BAND SCPC EQUIPMENT (SSB/SCPC)

Once the actual RTTY (teletype®) signals are located, the shift of the RTTY signal and the baud rate must be determined by stepping through the various speeds—50 baud, up to 1200 ASCII and Baudot Codes.

At this point it is necessary to become familiar with the operation of the RTTY demodulator (decoder) in order to receive copy from the various transponders. The operation manual covers the operation of the M-600 in detail. It takes time to learn how to tune and decode the material, with this equipment, however tuning this fascinating side of satellite dx'ing will fill many hours. Consult the table of contents for location of "list of satellites carrying data (SCPC)." See **Figure 4.37.**

ANOTHER NARROW BAND SCPC-FM/ SCPC/FDM

Again SSB/SCPC is fine for slow speed data, say up to 100 baud rate and human speech, however, it is not so good for music or high speed data. There is simply not sufficient "bandwidth" to support the greater range of baseband audio frequencies needed to support music, or the bandwidth required for fast data (110 to 1200 baud rate). A second technique for sending "narrow" band data via satellite was developed—FM/SCPC/FDM. This is typically a 60 kHz wide service; these 60 kHz wide channels are spaced across a transponder at roughly 200 kHz intervals. (Remember a SSB/SCPC signal can be as little as 4 kHz wide.)

If a transponder can be useful to a bandwidth of 36 MHz (that is 36,000 kHz), then in theory we could fit 600 channels if each were butted up against the adjacent channel(s). (36,000 divided by 60 equals 600.) Even at 200 kHz spacing, we could have 180 active channels. In the SCPC field, we will deal mainly with the FM/SCPC/FDM system, because these FM/ SCPC services have the greatest variety of information, educational and entertainment programming on board, and represent the largest area of growth.

MULTIPLEXING AND FREQUENCY DIVISION MULTIPLEXING (FDM)

In the early 1940's, the need arose for maximizing the amount of information that a particular channel could carry. This was crucial at that time due to the expense of laying underwater and underground cables. The telephone and communications companies began using a system called multiplexing which allowed one or more individual signals to be carried over a single cable, carrier, or other transmission medium. A technique was developed to allow many channels to be carried over a single medium. This is called frequency division multiplexing (FDM). Since the early days, this technique has been refined to allow

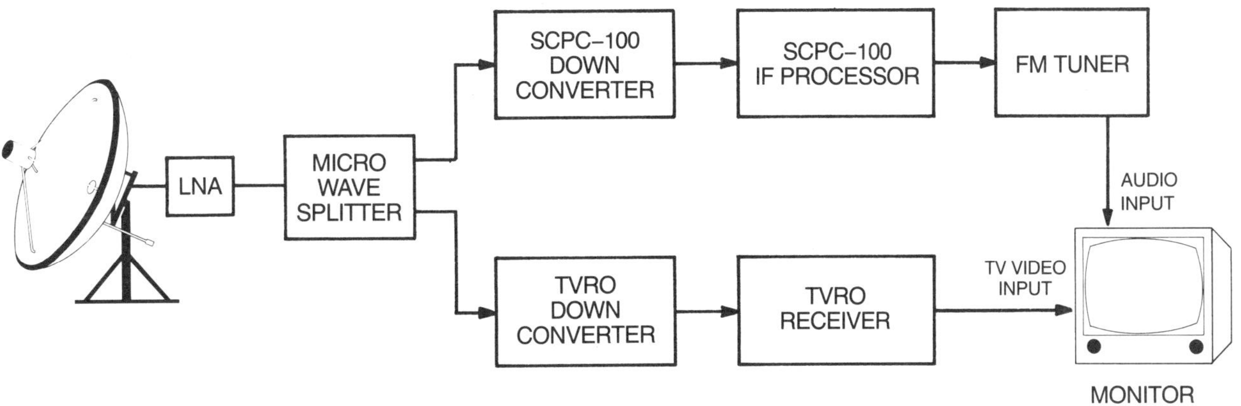

Figure 4.32 *SCPC method of audio recovery for international satellites.*

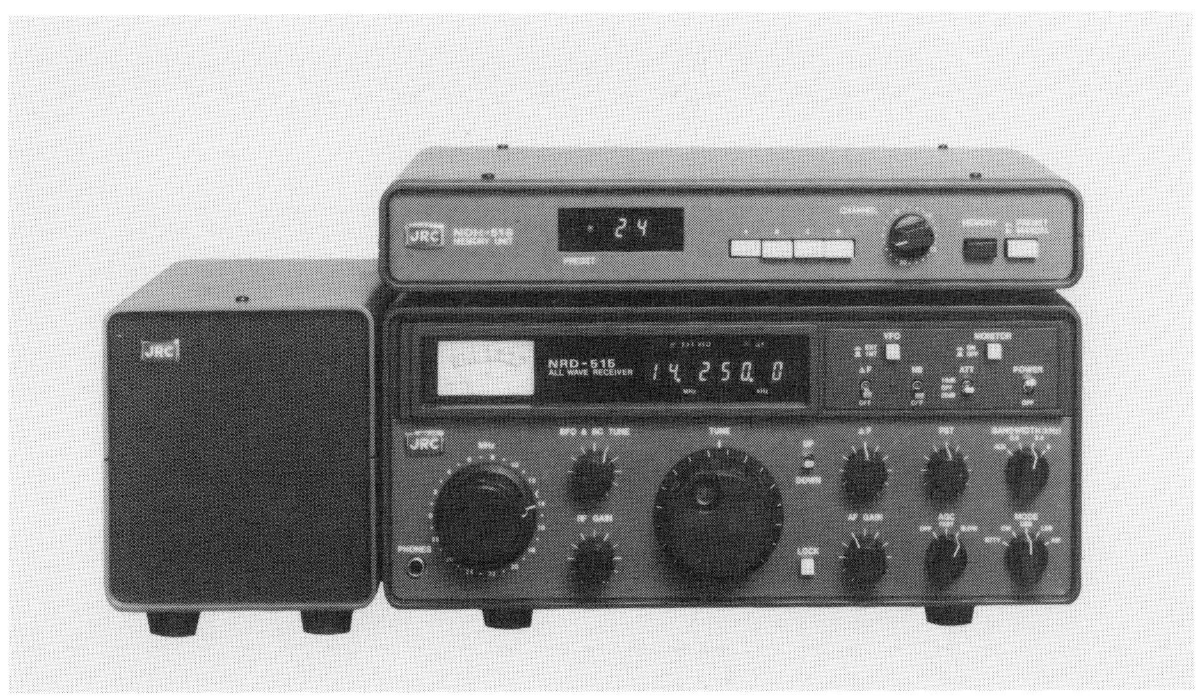

Figure 4.33 *SSB–General coverage .1 to 30 MHz receiver by Japan Radio—JRC–NRD–515.*

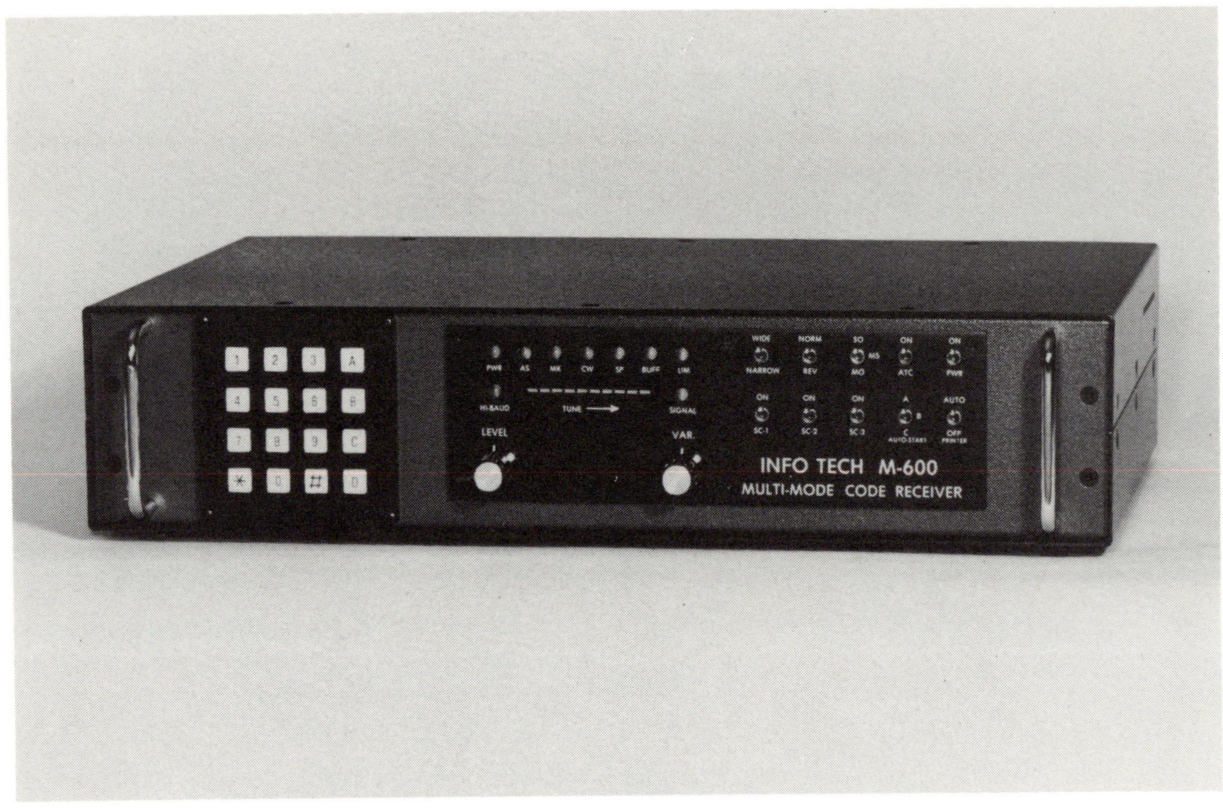

Figure 4.34 *M–600 RTTY teletype demodulator for decoding data.*

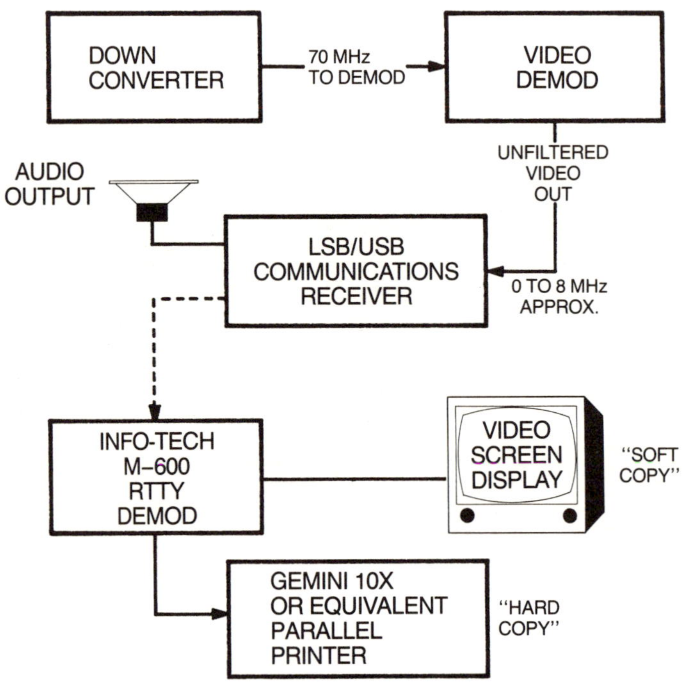

Figure 4.35 *Basic SSB/SCPC low baud rate data system with printer.*

Figure 4.36 *Data receiving equipment for SSB/FDM–SCPC format.*

Figure 4.37 *Complete satellite receiving equipment—video, all forms of data, facsimile, all forms of audio.*

many thousands of voice and data channels to be transmitted over a single cable, microwave link or satellite transponder, or part of the satellite transponder.

The most common system uses a set scheme of numbering of the frequencies used into channels. This is accomplished by accepted standard spacing, plus other elements for data use such as teleprinter shifts, etc., The International Telecommunications Union (ITU) set these standards with its consultive committee known as the CCITT. These specific frequency spacing numbering systems are used today in high frequency radioteletype work, cable teleprinter and on the teleprinter circuits and channels on satellite. The teleprinter data is sent in channel, twelve of which are often referred to as a group. When using an extremely broadband media link such as satellite transponders or microwave links, these groups are combined into five groups called a Supergroup (approximately 50 data or voice channels), and these Supergroups are combined into Mastergroups. These broadband media links are a frequency spectrum within themselves.

In these multiplexed channels resides the many thousands of voice and data transmissions in which we are interested. FDM can be combined with many systems of transmission, data, voice, and facsimile forms. The study of FDM would fill a book on its own.

Time division multiplexing—is a technique used for transmitting many channels of information over a single carrier by dividing a time frame into slots, with one slot dedicated to a specific channel of data. TDM is intermittent in its nature, and is mainly used to transmit digital data and digitized analog signals. It is possible to code and decode the information on a TDM system by using:

Synchronous TDM—in this techinque, all time slots on both the transmit and receive side are referenced to a master reference pulse, and a precise time interval from the start of a frame determines the location of the desired channel. (The channel's location is determined by time.)

Asynchronous TDM—this system uses the time slot for a specific channel which can occur randomly, and is identified at the start of a specific channel by means of a code or clock burst occurring prior to the transmission of the data stream.

There are several time division multiplexing (TDM) methods used in video work; these are:

(1) Simultaneous transmission and reception of alternating pictures (STRAP). This method and system was developed and patented by CBS, Inc. This system allows transmission of two video signals over a single carrier by transmitting only one field of each of the two video signals, and then averaging and reproducing the missing field of each video signal from the two transmitted fields by a digital method.

(2) Data insertion on lines in the vertical blanking interval. This additional TDM system can be synchronous or asynchronous in that the individual channels of information are put on specific video lines. The data is preceded by both a clock run-in burst to synchronize the clock on the receive side, and coding to insure that proper data is being transmitted.

HOW TO SET UP TO RECEIVE FM/SCPC/FDM DATA SERVICES

Several of the nation's news services (UPI) transmit their radio (audio) and regular Baudot teleprinter news service on Westar 3 transponder 1. These services are contained in the many groups of channels that were covered in the sections on multiplexing and frequency division multiplexing in this chapter. These services are using the FM/single channel per carrier/frequency division multiplex system. See **Figure 4.38.**

The system requires a small dish (6 to 8 foot diameter) LNA, down converter from TVRO receiver, a stable FM receiver capable of tuning 65 MHz to 85 MHz, a stable general coverage communications receiver that will tune from 6 kHz to 18 kHz with provisions for upper and lower sideband mode, a radioteletype R demodulator (M-600 unit) containing video charter generator, plus printer output to drive a printer for hard copy.

Each UPI circuit contains the wire service from many states. The services are sent at 60 and 100 words per minute (45 and 75 baud rate). On the TVRO systems that have a 70 MHz IF, the UPI channels will be found at approximately 75 MHz on the FM receiver. After this 75 MHz FM signal is demodulated to baseband audio, you will have a composite signal of UPI voice or radio feeds that come on, on the hour and half hour, plus several 25 channel frequency division multiplex (FDM) signals. These signals should be found at the center of the voice frequencies of 8.25 kHz and 12.25 kHz plus or minus .10 kHz. This total composite signal, containing voice and data signals, is fed into the antenna input of the communications receiver tuned to 8.25 kHz or 12.25 kHz on the upper sideband setting of the mode switch. See **Figure 4.39.**

The fine tuning of the SCPC data channels must be done with the fine tuning control or RIT control on the communications (SSB) receiver, in order to step through the 25 or 50 FDM channels. This set of channels is transmitted at either 45 baud rate (60 wpm) or 75 baud rate (100 wpm) speed.

The M-600 radioteletype® unit must be set at the proper RTTY shift, which is 170 Hz and at the correct baud rate (words per minute) receive speed. The phasing switch is placed in the normal phasing of the signal; if copy does not come up, set the normal/reverse switch to reverse setting.

It is imperative that you become proficient in the use of the M-600 on HF radio before attempting its use on the satellite data circuits.

Reception of this and other FDM signals requires a satellite receiver with a very stable down converter to tune this 60 kHz section of the total 30 MHz wide down converted spectrum.

There are many more FDM channels located on *Westar 3 transponder 1,* plus hundreds more in the satellite area; believe me you could spend months tuning the many transponders listed and never be able to cover them all. Satellite and transponders with data channels are listed in chapter *1*.

WHAT IS TELETYPE® (RTTY)?

The basic narrow band receiver (whether FM or SSB) produces an audio "message" to you via the speaker. That audio message may make sense to you (i.e., be intelligible), or, it may sound like gibberish. The gibberish, if it has the sound of "musical bells" or tones, is probably some form of radioteletype. Radioteletype (RTTY) is the international teleprinter system, which allows a text message made up of words and numbers to be transmitted from one location to another. A teletype machine may be made by one of several firms and the name Teletype® should always be followed by the notation ® since it connotes a registered trademark of the Teletype Corporation. We will use the word so often here that we'll simply make that notation from time to time to keep the people at the Teletype Corporation happy!

Teletype is a system of codes, either something called the Baudot Code or the ASCII Code. It is akin to (but not similar to) Morse Code. Each letter, number and punctuation mark has its own code, and when the operator strikes the key for T on the machine, that code is transmitted. It has a "musical sound," as do all teletype characters. The code is transmitted, and then received by a gadget called a "teletype demodulator." Since these are audio tones, you interconnect your teletype demodulator to your satellite (audio) receiver, so the tones you hear are fed into the teletype demodulator. Then the demodulator decodes the coded tones and supplies some type of output which will display the letters and numbers (words, et. al.) for you.

There are two ways to display this. One is the "soft display," which means that your teletype demodulator has a video output which is compatible with your video monitor or home computer monitor. Text appears on the screen, line by line, and "scrolls" or moves upwards as new lines come in. The current line always appears on the bottom of a full screen (although the screen will begin filling at the top and fill down until full), and then as each new line "comes in," the top line moves off the screen. The disadvantage to this should be obvious; you can only retain as much data as your screen will hold, and once a line flips off the top of the page you have lost it forever. Still, it is an inexpensive way to enjoy these transmissions and much of what you receive is not worth saving anyhow.

Another approach is to connect the output of the teletype demodulator to a printer, such as the popular Gemini 10X printer found in the $300 range at many discount computer stores. If your teletype demodulator has a "parallel output" port, it can then be connected directly to your parallel input port on your "hard copy" printer. Now you can copy and keep permanent records of what you receive for as long as your box of paper holds out!

Some of the transmission speeds are moderately fast—100 plus words per minute. Some are even faster, up to a thousand words a minute or so. Speed reading is impossible; you have to "save it" to study it later. Most, however, lumber along in the 100 word per minute and down region, and you can read and stay up with this without much difficulty. If you have a printer, you can save just those portions you wish, such as the stock market quotations or the Anik delivered CBC French Language news to impress your neighbor who speaks French.

Virtually all of the major world news and commodity and commercial money markets use teletype to send data. Most of these use satellite relay, and if you know when and where to look you can find virtually anything happening in the world, first hand, as it happens. We will note that one of the very best, and most reasonably priced, teletype demodulators on the market today is manufactured by Info-Tech (Model M–600, from Info-Tech, available from Universal Electronics, Inc., 4555 Groves Road, Suite 3, Columbus, Ohio 43232.

For a complete study on radioteletype for high frequency radio and satellite data work, you should obtain a copy of a new book devoted to radioteletype, "RTTY Today." This new book covers all phases of RTTY and computers in RTTY work. Cost is $8.95, plus $1.75 for shipping and handling, from Universal Electronics, Inc., address above.

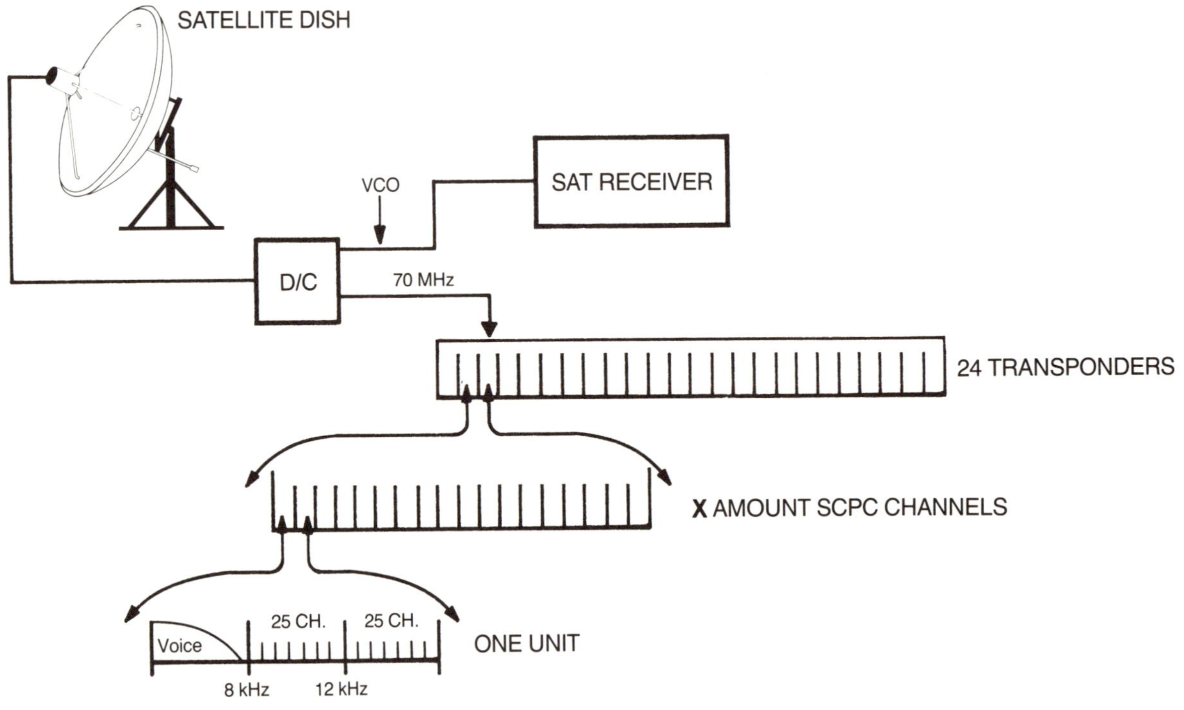

Figure 4.38 *Frequency division multiplex plan.*

A USER FRIENDLY FM/SCPC RECEIVER—

A typical consumer/home user FM/SCPC receiver should have the following features and uses:

A) The block down conversion approach, employed in the commercial version, is now optionally available with a standard single conversion down converter. This makes the system more compatible with the typical home TVRO installation.

B) The rack styled receiver is now in a stand-alone, consumer-styled package (which we have not seen as this is written).

C) The professional-grade receiver assumes the operator of the system will be a technical-type person. A pair of initial set-up adjustments under the top lid of the receiver have been brought out front. A meter which is used in the professional-grade receiver to properly tune in the individual carriers has been replaced with an easy-to-read and understand LCD (liquid crystal display) four digit display which serves two functions:

   1) It tells the user when he is properly "center tuned" on the individual FM channel, and;

   2) It tells the user where in the full transponder "band" he is tuned. More about this shortly.

D) A pair of audio outputs are available—one is a two-watt audio amplifier designed to feed directly into an external speaker (supplied with the unit), and the other is a 600 ohm "balanced" output to allow the user to feed additional peripheral equipment for the "copying" of radioteletype R text, news services, and so on.

E) And finally, and hardly the least significant, the package has to be in the consumer format so that for $995 (suggested retail), the home user can add an entirely new dimension in high quality news, entertainment and information, "audio programming" to his TVRO, or, his stand along "ARO."

Now it happens that many of the users of this particular transmission format are getting by very nicely with some extremely small dish systems. Muzak (one of the many available here), for example, goes typically to four foot (and even smaller) dishes on a commercial basis! That means that with the consumer-style receiver, you could be out there selling and installing high grade satellite audio reception systems with some very tiny dishes.

OPERATION

As noted, the professional-grade receiver may not be desirable for the average consumer user; unless this was one of those installations where the dealer was going to put the unit in, tune in the receiver to the proper audio or data service channel, and then walk away leaving the receiver in one dial position. The average consumer is going to treat the system just as he treats his AM or FM radio; there is a "dial full" of stations there, some carrying news, some carrying sports, some weather and some music. He is going to want to follow the Minnesota Twins baseball team in the afternoon (using the Minnesota Satellite Radio Network), and attend a concert from Carnegie Hall on NPR radio that evening. The professional-grade receiver requires that the user "remember" where each station is without benefit of an accurate frequency readout dial.

The consumer model has a four digit meter on the front panel, and as you turn the tuning knob, the liquid crystal displayed numbers change. Those numbers relate to the frequency of the signal being tuned in, and the user is supplied with a series of pre-printed cards which slide into a holder on the front of the receiver. There is a card for each transponder (i.e., TR1, W4, TR3, W4), and space for the user to "log in" the digital meter reading that corresponds to each station tuning position. The same meter is also used for two other purposes.

Because of the 30 Hz waveform energy that some (all are required; many do not comply with FCC regulations) signals carry, there are two initial set-up adjustments which the dealer must make when installing the receiver (this assumes the dealer will install it; that the consumer will not be asked to do it, although in truth he could). The meter makes these two adjustments painless. Then, after the adjustments have been set, the user finds the meter handy for "center tuning" of the signal—just like you have on some of the TVRO video receivers, or the stereo signal processors, like those available from R.L. Drake and Arunta.

The transmissions from the special satellite feeds are "high fidelity," but not in stereo. Well, let's correct that! Some are in stereo (NPR, for example), but to create the full stereo effect, you would have to install a pair of these receivers so that both the L (left) and R (right) channels could be independently received and fed to a stereo amplifier system. Most of the stereo services (other than NPR) are found on the subcarrier format anyhow, which simply means that existing hardware in the marketplace is available for the stereo fan.

FUN TO OPERATE

Other than a dedicated terminal sold to a Muzak user, or to a local radio station or newspaper, the user of the sFMs package will quickly discover that there is as much "audio fun" on satellite, in the FM/SCPC mode, as there is "video fun." Virtually every major league baseball team now has a satellite connected radio network service; that means that you can be a Montreal fan in Los Angeles or a Los Angeles fan in Florida (or the Bahamas) and never miss a game. Reception is studio quality, and because the listener is "inside" the radio network feed, it is the audio equivalent to tuning in direct back feeds on video of sporting events.

All of the radio networks supply their affiliates with multiple newscasts per hour. Between Mutual, ABC, CBS and NBC, you are never more than a few minutes from a fresh radio newscast. State radio networks abound; you can stay up with Florida news in Maine or Minnesota news in Texas. Special news feeds for the Caribbean, Central and South America are up there. The Naval Observatory Time (accurate to a few parts in a million) is there. A "talking book" service for the blind is there.

Services are spread across W3, W4, and F1R primarily. Service is scheduled for F2R, and G2. Since the uplink or bird operators tend to bunch multiple users onto a single transponder for ease of switching and maximum transponder efficiency, there is not much of a "fishing expedition" required to locate those that are active.

CONFIGURING A SYSTEM

There are several ways to plan a system such as this; the best design will depend upon the user himself.

If the user is casual, and he is willing to share the FM/SCPC with the TV function of the dish, a single down converter used for the TVRO and ARO system would work; a switch or relay installed in the RG-59/U line coming from the down converter to the TVRO receiver would allow the user to switch off of the TV service and onto the ARO service. The down converter would be powered by either the ARO unit or the TVRO unit; that unit would be left on fulltime, regardless of whether the user was watching television or listening to satellite fed radio.

If the user was interested in joint TV and audio use, a pair of down converters with a 4 GHz two-way splitter at the dish, and parallel runs of RG-59/U would be installed. What is wrong with this approach is that when you are on Westar 4, TR 1 or 3, you would be limited to the horizontal transponders on W4 for TV at the same time (PBS is there).

BUILD A FM/SCPC SYSTEM FOR UNDER $100

The concept of extracting non-video information from the many Clarke orbit satellites pre-dates the use of video, since all of the early experiments were actually "narrow band" or "non-video" based. What is "new" is the tremendous proliferation of non-video material, much of it informational or educational in content, now being fed via satellite worldwide on a routine basis.

Ours has become an intensive "information based" society. We depend upon and often profit from the information which we find relevant to our business activities or lifestyles. The transfer of data, or information if you are not quite up to "computer speed" yet, is becoming as important as the creation of data. In some quarters, it has become even more important, and the satellite connection makes it all possible, worldwide, for exceedingly low cost.

Further proof of this can be found in the latest Intelsat proposals where they are now offering complete 12 GHz receive-only "data terminals," using two foot dishes for around $2500 (U.S.). They claim that if the terminal is used within the hotter boresight footprint of the latest Intelsat V birds, such a two foot dish propped up on a table and pointing "through" a glass window will spit out high speed data information relayed via satellite. What all of this has in common is that virtually all data of interest is transmitted, using one or another "narrow band" methods.

Our TVRO (video) terminals are all wideband systems; that is, we require between 15 and 30 MHz of "information" to recreate on our screens, a television picture (and the companion sound). On the other hand, a narrow band system requires only a fraction of this "bandwidth window" (measured in megahertz or MHz) to transmit all of the contents of the message. The message contents may be audio (voice or music or both), or it may be some type of computer data code such as ASCII (pronounced ASK-KEY). Our concern at this time is not so much how you make intelligence out of the various codes, as it is in equipping you to receive the basic transmissions in the narrow band format.

We will show you here a couple of ways you can explore the world of FM/SCPC. For well under $100, you can be enjoying FM/SCPC reception this evening.

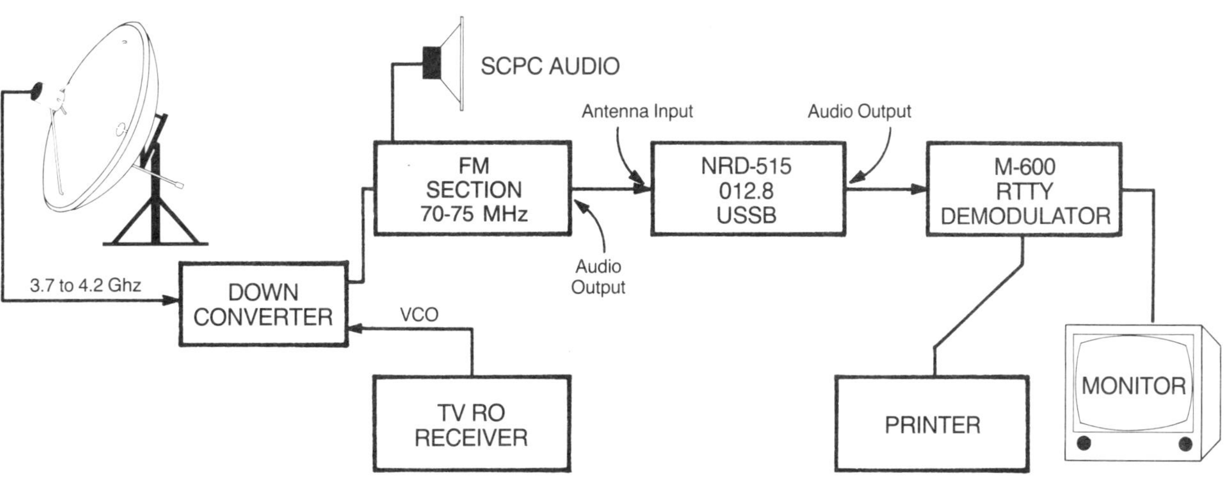

Figure 4.39 *Equipment diagram for receiving FDM/SCPC news services.*

Naturally there are some trade-offs involved if you are going to "sneak into" FM/SCPC for less than the cost of a first rate FM narrow band demodulator. The primary trade-offs are as follows:

1) You will have to tune the tuning dial far more carefully, even slowly, to separate the many station services listed here and found on satellite;
2) You will probably have to "touch up" the tuning of the special adapter we will be describing, every 15 minutes or so, to compensate for the "drift" of the adapter and the balance of the system;
3) You may not be able to "separate" (as in pull-apart) all of the services when there are two or more, which are located fairly close together within a transponder.

Still, for less than $100 (and as little as $25), you will in short order be listening to these wonderful and elusive narrow band audio type services and you will be well on the way to an entire new level of "satellite exploration."

BASICS FIRST

Most of us has a good grasp that when a TV signal is transmitted via satellite (relay), the video carrier has a certain frequency and that the video information (detail, color, et. al.) "spread out" from that carrier frequency. This is the primary reason why a video signal utilizes so much of a transponder; there is so much information being transmitted, that it cannot all be "crammed into" a tiny or narrow bandwidth. We are also aware that an audio signal, such as a subcarrier, occupies a far smaller "width" in the spectrum because there is far less information present. In fact, many audio subcarriers can be crammed into a relatively tiny portion of the spectrum, as anyone with a subcarrier tuning system can attest to by tuning across the subcarrier region of F3R's TR3 (with WGN video). As many as 15 separate audio subcarriers have been found there at one time, all without interfering with one another nor ostensibly with the video carrier they companion to.

Imagine for a moment that you had a transponder filled with just audio subcarriers. If they can have 15 (or more!) share the video transponder with WGN, how many do you suppose could be there if they took away the WGN video and filled the WGN video area with just audio subcarriers? Hundreds. And that basically, is what FM/SCPC does. This is not how it works however, because the audio subcarrier system depends totally upon the presence of the video carrier in the first place. We cannot simply "take away" the video carrier, and fill in the space vacated with audio-subcarriers because the SUB carriers are "sub" (as in "attached to") to something; that something is the video carrier. The video carrier is the vehicle which the audio subcarriers ride into space and through the satellite.

So if the system designers want to stack hundreds of "subcarrier-like" audio (or other narrow band) signals into a single transponder, without a video carrier present for them to "ride along on," they must use a technique which allows each individual "subcarrier" to stand alone, all by itself. This is not such a bad deal, since it means that you can transmit one of these narrow band channels from an uplink in Charlotte (N.C.), another from an uplink in Minneapolis, a third from an uplink in Dallas, and so on; each stands alone and each occupies a tiny portion of the spectrum within a transponder.

The bad part, for you, is that your TVRO receiver depends upon the video carrier's presence to "lock onto" a signal and stabilize your receiving system. That transponder you tune in is actually a "signal reference" which the receiver latches to, to insure that your receiver does not "drift away from" the signal. If the video is present, and the TVRO receiver has "locked onto" the video signal and is holding it stable, then the audio subcarriers that ride along with the video carrier will also be stable. In short, you can set your receiver to F3R's TR3, it will lock up on WGN, and then with your built-in audio subcarrier detector or an outboard unit from Arunta, Drake, Maspro (etc.), you can tune in any TR3 audio subcarrier you wish and it will also stay "locked in."

When you establish a system to tune in an FM/SCPC signal, on a transponder where there is no video signal present, your FM/SCPC receiving system has nothing to "lock onto;" the FM narrow band signals are very complicated to "lock to" and so far there have been no "low cost" techniques developed to overcome this problem. And in fact, that is a significant part of the cost in the dedicated FM/SCPC receiver; the "locking system" that allows you to set the receiver tuning dial to one particular FM/SCPC signal, and come back hours later and still find it "tuned in."

The two techniques to be described here are minus this "lock" function. How stable your reception will be will depend largely on the stability of your TVRO receiver's down converter. To the best of our knowledge, no exhaustive comparison testing has been done between different brands of down converters to determine which of those presently in the marketplace are "stable" and which are not, given no video carrier being present to "lock onto." Perhaps as a few hundred people try the system we are about to describe, we will get reader feedback on which down converters seem to function best in this "service."

TUNING THE IF

Your down converter, whether a stand alone unit at the antenna (i.e., Drake), or an LNC package (i.e., USS/Maspro) will send a 70 MHz (centered) IF signal indoors through the connecting (typically RG-59/U) cable. We will place something at the indoor end of that cable to allow us to feed the 70 MHz (centered) signal into a secondary tuning unit; one which will allow us to tune in the FM/SCPC stuff. We will spend less than $100 to do this.

In **Figure 4.40** we have a typical system with an addition—the antenna plus LNA feeds signal to the antenna mounted down converter. If your system uses an LNC, they would both be together at the feed. The important point is that you have a 70 MHz signal coming indoors in the cable. Note that this diagram shows a system which does not send DC operating voltage for the LNA (LNC) and down converter through the same cable. There are still some receiver systems in use that send the outdoor voltages through a separate line or set of wires.

Indoors, we install a two-way HYBRID splitter. The input faces toward the 70 MHz input side or the antenna. We have two-outputs: one of which we connect directly to the TVRO receiver, and the other we connect to the "FM/SCPC Interface Converter."

The "Interface Converter" is nothing more than a $24.95 (last price shown) Radio Shack (Automotive) VHF-TV Sound and Weather Converter ND 12-1354. This is a device that tunes two bands, the VHF (TV) low band and the VHF (TV) high band. The low band tunes from approximately 50 MHz to approximately 90 MHz. This means it tunes through our standard TVRO 70 MHz IF.

Remember that a 70 MHz IF is only centered on 70 MHz; it actually is wider than this, covering from 59 MHz to 81 MHz for a TVRO receiver with a 22 MHz wide IF and so on up to as much as 40 MHz wide. As the 70 MHz IF signal comes indoors from the down converter, it will be plenty wide enough to cover the entire 50-90 MHz spectrum we have an interest in here. Also remember, that a full satellite transponder is 40 MHz wide, and the center of that 40 MHz width falls at 70 MHz while the low end falls at 50 MHz (70-20) and the top falls at 90 MHz (70 + 20). Yup! 20 plus 20 still equals 40. If we have a receiving device that will tune across the 50-90 MHz band, we have an instrument which will allow us to tune in the individual FM/SCPC carriers! We'll return to this shortly.

In **Figure 4.41,** we have a system which also sends the DC operating voltage to the LNA/down converter/LNC via the same coaxial cable. That means we have 70 MHz (centered) signals coming indoors from outside, and we have a DC operating voltage coming outdoors from inside. One good way to blow up everything would be to connect the DC operating voltage intended for the outdoor equipment directly to the Radio Shack (or equivalent) "Interface Converter." So we need a system to keep the DC voltage going to the outdoor equipment, but not to the "Interfacing Converter." The device is called a "directional coupler," and you will find them at the nearest MATV type store or wholesale house. You want one that is "10 dB down," or in other words, one that attenuates the incoming signal going to the "tap" output side by 10 dB while allowing the main signal to pass through essentially unattenuated to the TVRO receiver proper.

This directional coupler allows the TVRO IF signal from the down converter to pass directly to the TVRO demodulator; it allows the DC operating voltage to trace the reverse path back to the down converter (etc.) outside, and it isolates or prevents any of the DC power from getting into the "Interface Converter." Cost? Under $5.00.

Now the Radio Shack model 12-1354 is intended for automotive use; it is supposed to allow you to sit in your car and tune in the audio portion of local TV broadcasts. That means it is an "FM receiver" of some sort, since your TV audio broadcasts are FM. Here is how it works.

The FM signals, tuned in from the local channel 2, 3, 4, 5 or 6 stations, are down converted inside of the box to an FM-IF. Then the signals are amplified, and detected. That is, they are turned into audio. But

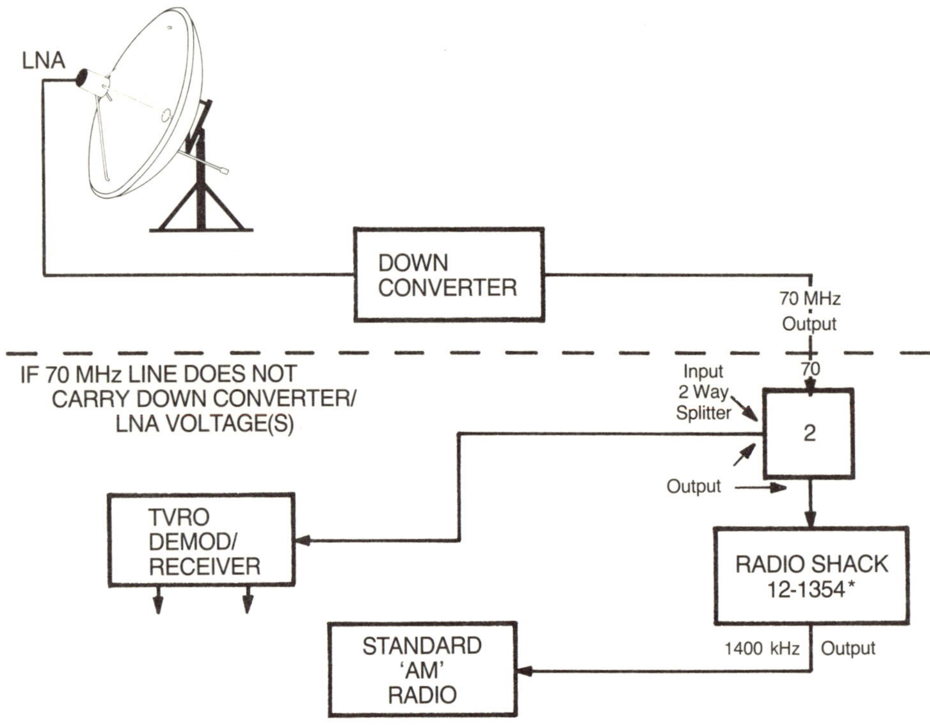

Figure 4.40 *Quick solution to tuning-in FM/SCPC signals off of satellite.*

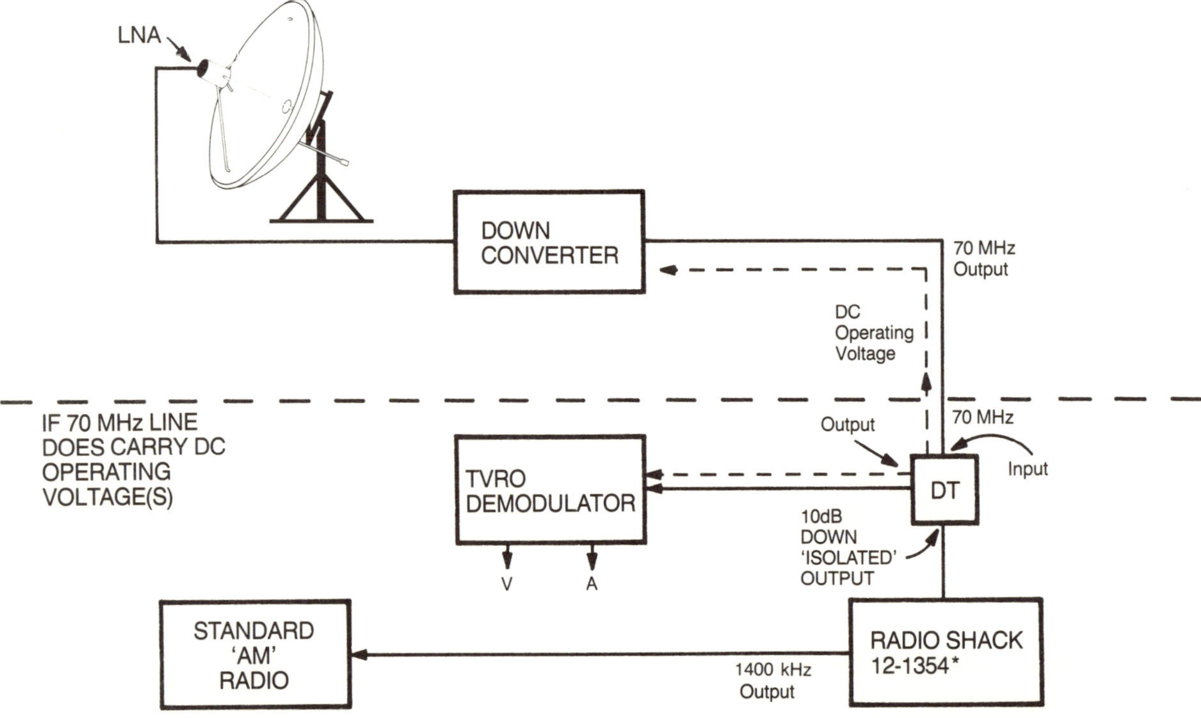

Figure 4.41 *DC power between demod and down converter must be 'blocked' so as to not go to FM/SCPC converter-receiver.*

the object of the 12–1354 is to allow you to tune in the FM–TV audio on an AM radio. So they take the audio and connect it to a low power AM transmitter inside of the box. Yup, that sounds just like a TV modulator built inside of a TVRO receiver!

Then they tell you to connect the output of the 12–1354 box to a standard AM radio; any standard AM radio. Tune the AM radio to around 1400 on the dial (1400 kHz or 1.4 MHz) and there you will find the carrier from the 12–1354 box. If you have a local station near 1400, they give you a slug to tune, to move the output AM signal coming from the box away from 1400 so you can find a clear, local, spot on your AM radio dial.

To operate the system, you tune the master TV tuning dial on the 12–1354 to the correct operating frequency of your desired FM/SCPC signal. The audio from that signal will appear on your AM radio at or near 1400 kHz. That's it. Not very complicated although since the 12–1354 was intended for automobile use, you will have to find 12 VDC from your TRVO receiver as well (such as the often-provided optional voltage line at the back of your receiver; caution . . . be sure the voltage here does not exceed 13.8 VDC or you could blow up the Interface Converter!). Complete installation instructions for the 12–1354 are included in the box.

There is one other possibility here if you are into the big bucks (such as spending $69.95 or so), and you don't wish to fool with finding 12 VDC for the 12–1354. Go to the same Radio Shack store, or Sears Roebuck or some other 'high quality' electronics store and ask to see their "Multi-Band/TV Sound" receiver. Most stores now offer you a portable receiver with built-in power supply, speaker and batteries which tunes directly the TV band signals for channels 2–6 (that's the band you are interested in), and channels 7–13. One caution here, you will need to purchase a high quality receiver with a built-in connection for an outdoor antenna. Your "outdoor antenna" will be the piece of RG–59/U that comes from the directional coupler or signal splitter output side, and if you want to be very pure about it, put a small matching transformer on the end of the cable before connecting the coax to the typically 300 ohm antenna terminals on the "multi-band" receiver.

OPERATION

The Radio Shack 12–1354 or the Multi-Band receiver(s) is designed to recover FM signals that are typically around 150/200 kHz wide. The narrow band FM/SCPC signals on the other hand are closer to 60 kHz "wide." That means you will be getting lots of extra fidelity, but you also may be getting extra noise with this system. And if you are tuning in an FM/SCPC signal which is close in frequency to others on the same transponder, you may experience some problems separating the two (or three) signals. But then again, the price is right.

Shown here is a diagram which perhaps will assist you in determining how my listings correlate to where you should tune in your receiver dial to find the same signals I am reporting. The top line is the (as marked on the unit) Radio Shack 12–1354 tuning dial. The next line down is your 70 MHz IF, which as we know really is a 50–90 MHz IF, as the signal comes indoors from the down converter or LNC. See **Figure 4.42.** The third line down is my own "dial," the digital frequency tuning system we use with our receiver system. If you see a listing of some service you want to tune in from our tables, and the listing says 25, that means you would tune your 1354 tuner to the region just below where the 1354 has Channel 2 marked. (Note: The dial calibration on the 1354 units we have tried is not great, so you may experience some variation here).

The bottom line is the reference for you; a typical transponder at its original "GHz" frequency, as that would compare with the above three reference lines. (Note: A multi-band receiver, tuning in low band TV audio, would have the same dial calibration as the top line since the 12–1354 and the TV audio tuner/receiver are essentially the same type of gadget. The TV audio receiver is simply more complete, and ultimately perhaps a tad higher in quality).

There are alternate methods to accomplishing this "trick." Some people have taken high quality FM tuners (tuning 88–108 MHz) and they have modified the oscillator so that the tuner now tunes 60–80 MHz (that's part of the "band" anyhow). You also must modify the input RF stages of the tuner as well, when you do this, since they were designed to tune 88–108 MHz, not 60–80 MHz. Overall, the quickest cheapest way to get into FM/SCPC is as outlined here. You will discover, as we have, that there are considerable performance variations between different down converters in this service (the Drake seems to work very well).

The following listings show you where you can find the (typically) 60 kHz wide FM/SCPC services with a narrow band FM receiving system or, the "experimental" systems described. Not all services will be found operational 24 hours per day and many maintain irregular hours within the 24 hour day segment. Our designations are as follows:

1) N = known to be less than full time
2) MN = may not be transmitting full time
3) S = schedule times, examples given
4) SE = special events, such as baseball game coverage only

Many transponders have one or more unmodulated carriers present, i.e., these carriers are always (or nearly always) "up," but no audio or data has ever been noted. Most services that are not full time will keep their carriers operational between the actual "broadcasts." When you tune across an unmodulated carrier, you simply find receiver "quieting" (i.e., the static noise disappears and the sound from the speaker becomes "quiet"), which will abruptly turn into programming at the scheduled time. An example of the latter would be Mutual Radio News which may send news at 00, 15, 30 and 45 minutes (past the hour) and if the newscast lasts five minutes, than that particular audio channel SCPC service would be active from 00 to 05, 15 to 20, 30 to 35, and 45 to 50 minutes.

WHERE TO FIND SCPC SERVICES

Listed below are considerable SCPC locations and services, these are only a few of the hundreds of special audio networks which are available on the satellites. Not all services will be operational 24 hours a day, the designations are as follows:

(1) N = Known to be less than full time
(2) MN = May not be transmitting full time
(3) S = Schedule times
(4) SE = Special events, sports coverage only
(5) BLANK = Active most of the time

Satellite	Transponder	Designation	Service
W4	1(H)		**Mutual Radio**
			Mutual Radio
			Mutual Radio
	2(V)		National Public Radio
			National Public Radio
			National Public Radio
			Handicap/NPR
			Mutual Radio
			Mutual Radio
			Georgia Radio/News Net
			Georgia Radio WGST
			'Elevator Music'
			IMS News
			Alabama Net/News
			National Public Radio
			Mutual Radio

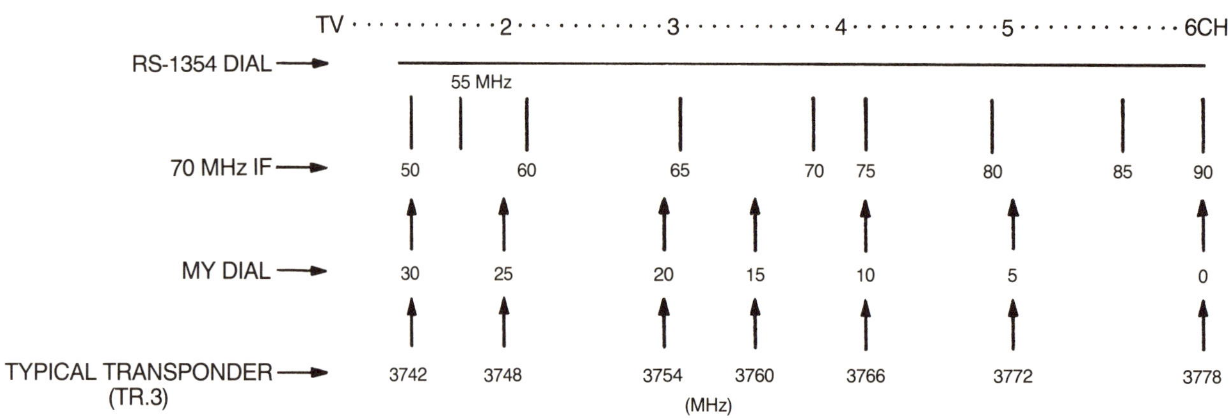

Figure 4.42 *How to relate your tuning dial to the receiver system you are using. RS–1354 dial must be precisely set for each individual signal (top line). TV audio receiver tunes directly on '70 MHz IF' or second line. SCPC receiver tunes line three. Line four shows typical 40 MHz wide transponder and relation of frequency to top three 'scales'.*

Satellite	Transponder	Designation	Service
W3	3(H)	MN	U.S. Navy/Master Clock
			Minnesota Net/News
			Alabama Info Net
			Mutual Radio
			Mutual Advisory
			Minnesota Net/News
			Mutual Radio
			National Public Radio
			U.S. Navy/Master Clock
			In Touch Network
		SE	**Atlanta Braves Net**
			In Touch Net
			National Public Radio
			National Public Radio
			National Public Radio
			National Public Radio
			National Public Radio
			National Public Radio
		N	Texaco Opera Net/NPR
		N	Texaco Opera Net/NPR
			Mutual Radio
		MN	Mutual Advisory
			Georgia Radio Net
			Georgia Radio Net
			'Elevator Music'
	4(V)	MN	U.S. Navy/Master Clock
			Alabama Radio Net
			In Touch Net
			National Public Radio
			Mutual Radio
			Minnesota Radio Net
			Mutual Radio Net
			Minnesota Radio Net
			Mutual Radio
			National Public Radio
		MN	U.S. Navy/Master Clock
			In Touch Net
			Georgia Radio Net
			In Touch Net
			National Public Radio
			National Public Radio
	1(H)		UPI News
			AP News
		S On Hour	Wallstreet News
	2(H)		New 97 Radio
			Kansas Radio Net
			ABC Radio News
			Alabama Radio Net
			Oklahoma News Net
			Florida News Net
			Transtar Radio Net
			Transtar Radio Net
			UPI Radio News
	3(H)		KTRZ/Portland (Z100)
			(religion radio)
			N. Carolina News Net
			WRAL (Raleigh)
			WFBR (Baltimore)
			Mississippi Radio Net
			Louisiana Radio Net
			S. Carolina News Net
			Brownfield Radio Net
			Texas State Radio Net
			Louisiana Radio Net
		SE	K.C. Royals BB Net
			Kansas Info Net
			Oklahoma Radio Net
			Georgia Radio Net
			UPI Radio
			Transtar Radio Net
	4(H)		KTRZ Portland (Z100)
			'Elevator Music'
			Arkansas Radio Net
			N. Carolina Radio Net
			Mississippi Radio Net
			Louisiana Radio Net
			UPI Total Radio
	6(H)		'Elevator Music'
	7(H)		BBN Religion
			BBN Religion
	8(H)		KBC Talk Net
			BBN Religion
			BBN Religion

A SUMMARY OF SCPC

A brief summary to clear up any misunderstanding regarding the SCPC signals. We know it is not a carrier by itself, not modulated onto the main carrier, and consequently not available at the output of the receivers discriminator, so what is it? We first must understand a subcarrier—a subcarrier might better be called an *"attached carrier"* because it cannot stand by itself. The only pure carrier in a satellite TVRO system is the video carrier; it is generated by itself, modulated with the video information, and it contains all of the elements required to get from 'here' (the uplink site) to "there" (your downlink site). Even the program audio (the sound that is the audio for the picture) is sent on a 'subcarrier' (or attached-carrier). A subcarrier cannot travel from "here" to "there" on its own. It is introduced into the uplink part of the system by attaching it to the video carrier. If you take away the video carrier, you lose the subcarrier as well. Multiple subcarriers, the type you can tune in with a Drake SA-24, Arunta SSP-318 and so on, or with a TVRO receiver with subcarrier tuning built-in, are all *"attached carriers."* Take away the video, and they all go away: even if their audio material seems to be totally unrelated to the 'content' of the picture carrier. Program wise they are separate; technically, they are impossible without the video carrier.

When we speak of a "36 MHz bandwidth," we are talking about the full, *usable,* bandwidth at the RF carrier frequency of the up or down link. Transponder 1, for example, extends from (typically) 3702 to 3738 MHz. Less accurate charts show it to be 3700 to 3720 or 3.7 to 3.72 GHz. That 36 MHz bandwidth persists or stays with us from the uplink transmitter, through the satellite, to your down converter and through your receiver IF. A 70 MHz IF? It has a 70 MHz "center channel frequency" but it actually is quite wide: from 52 MHz to 88 MHz if it is capable of passing and processing the full channel or transponder width.

This "70 MHz signal" is processed through the "IF" stages to the demodulator (same as discriminator). There it is turned from a modulated carrier wave signal that is 36 MHz wide to a "baseband" signal that is, at the very-very most, around 10.7 MHz wide. That 10.7 MHz wide "signal" can be thought of as the "final" or "last conversion stage" in the receiver. The information that falls between 0 MHz and 10.7 MHz, coming out of the discriminator/demodulator, is where we find our video (picture) information. Attach that 0 to 10.7 MHz wide signal to a video picture tube (i.e. monitor) and you will have television pictures displayed!

So what "was" 36 MHz wide is now "no-more-than-10.7 MHz" wide, *after the discriminator;* a form of "bandwidth compression" or "frequency crunching." However, the actual video picture information will occupy no more than 4.2 to 4.5 MHz *(from 0 MHz to 4.2/4.5 MHz)* of that as-wide-as 10.7 MHz signal. That leaves a considerable amount of frequency space to tack on "attached signals." The subcarriers attach in this region, usually no lower than 5.0 MHz and no higher than 8.1 MHz. (In the *real world,* the original transponder is usually not fully used to its full 36 MHz width—30 MHz is more normal—so if we *lose part of* the overall 'width' we will *also lose part of* the 0–10.7 MHz 'width' as well.)

When you tune in a transponder that does *not have* a TV signal on it, they may be using it for other forms of communication (TV being but one 'form'). One of these forms is called SCPC (signal *channel per* carrier, *or,* single *carrier per* channel, depending upon which 'school' you went to!). SCPC can be tuned in, as you suggest, by connecting the baseband or (normally) video output from your TVRO receiver to a communications receiver that tunes the frequency range from 0 to say 8 MHz. If there is *no video* there, and *there is* SCPC use of that transponder, you will tune in these SCPC signals on the communications receiver.

We call these *SSB/SCPC* because they are transmitted using a modulation format known as single sideband (SSB for short). There is another type of SCPC as well; using a frequency modulation rather than a single sideband modulation format. All of the services we have written about, to date here are *FM/SCPC* signals. Because they are FM, and not SSB, to tune them in requires an FM receiver that is capable of tuning *NOT* the 0 to 8 MHz frequency range but rather *the original* 70 MHz receiver range prior to the discriminator/demodulator.

Why is this?

The demodulator/discriminator is a *wide*band device. These are *"narrow* band" FM carriers. They are not 36 MHz wide (or 30 MHz wide) *each;* they are but a tiny fraction of that wide (typically 60 to 100 *kilo*hertz or kHz). If a wideband discriminator demodulates them, they are lost; gone. If a single transponder has two or 200 of them, they all get demodulated as a "lump," all together. You get no "audio" or intelligence out of that; only noise. Thus the requirement for the special FM demodulator receivers that tune through the 50–90 MHz (IF) region, to catch these signals *while they are still "RF carriers"* and before they get messed up in the wideband demodulator in our TVRO receiver.

WHAT IT TAKES To Tune It In:

1) Video: Wideband receiver, capable of 14 to 36 MHz bandwidth.
2) SSB/SCPC: Wideband down converter going to a wideband demodulator (30 MHz or better with an 8 MHz or better baseband output bandwidth) in turn going to a selectable sideband (SSB) upper/lower sideband communications receiver tuning 0 (.1) MHz to 8 MHz (+).
3) FM/SCPC: Wideband down converter with a 70 MHz (centered) IF connected to a 60/100 kHz bandwidth FM audio receiver that can be tuned *between* 50 and 90 MHz.
4) Subcarriers: Wideband (video) receiver capable of processing video in an IF from 14 to 46 MHz wide, a wideband video discriminator/demodulator capable of passing 0 to 8.0 MHz, feeding a 100/200/400/800 kHz wide FM detector that tunes from 5.0 to 8.1 MHz.

COMPANDING

A system of compressing the satellite signal at the transmission (uplink) side of the satellite system and transmitting this compressed signal, usually 2:1 or 3:1 compressed state, over the system. At the receive (downlink) side, this signal is received and run through a device to re-expand the signal to its original dynamic range or expanded state.

The basic system consists of a series of compressors and expanders, which allows the information to be condensed to one-half or one-third its original range.

Companding is used to increase the number of channels and traffic a 36 MHz transponder can handle. In use, a typical transponder's message handling can be raised from approximately 1,000 channels to over 2,000 channels using 2:1 companding, or from 1,000 channels to almost 3,000 channels using 3:1 companding. This method was first used by the military and was considered a top military secret for many years.

In addition, to increased traffic handling capabilities, companding can also improve the overall signal-to-noise ratios as much as 16 dB with 2:1 companding, and as much as 26 dB with 3:1 companding. Almost all forms of data and voice transmission can benefit from this system.

FDM/FM systems use the companding very successfully, along with many of the radio networks who use satellite distribution systems.

Figure 4.43

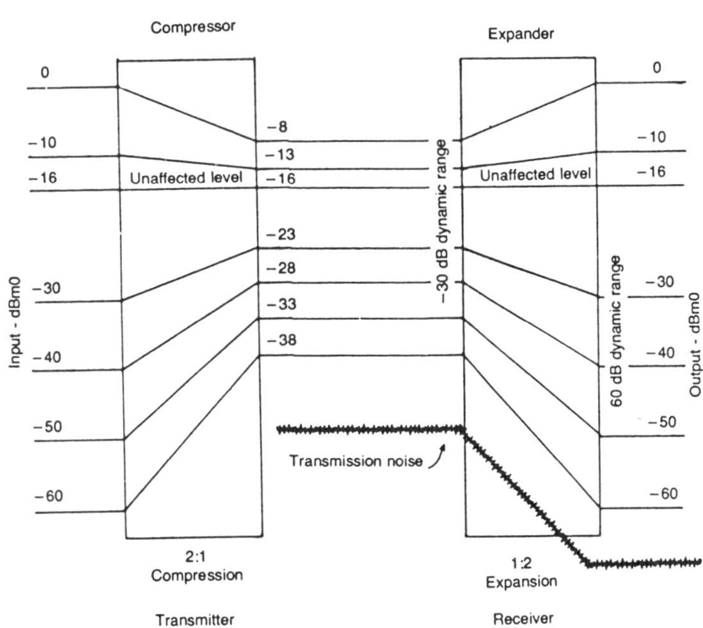

Figure 4.43 *"Companding" showing 2:1 and levels of compression and re-expansion to original levels.*
Courtesy Communication Consultants, Inc.

FREQUENCY LOCATIONS OF SCPC SERVICES

Listed below are satellite, transponder locations of many SCPC services. The first column lists the SCPC FM frequency of the service.

WESTAR 3, T1

FREQUENCY	SERVICE	COMPANDING	PRE-EMPHASIS	BANDWIDTH
62.9 MHz	Dow-Jones	2:1	75 μs	NB
64.4 "	Wold I	2:1	25 μs	NB
65.0 "	Wold II	2:1	25 μs	NB
65.6 "	Music Country	2:1	25 μs	NB
66.6 "	AP Radio	2:1	75 μs	MCPC
68.3 "	Data Stream			WB
71.9 "	UPI Data	2:1	75 μs	MCPC
75.0 "	UPI Audio	2:1	75 μs	WB

WESTAR 3, T2

FREQUENCY	SERVICE	COMPANDING	PRE-EMPHASIS	BANDWIDTH
52.0 MHz	Oklahoma News Net	2:1	75 μs	NB
52.6 "	Transtar	2:1	75 μs	WB-Stereo
52.8 "	Oklahoma News Net	2:1	75 μs	NB
53.0 "	Transtar	2:1	75 μs	WB-Stereo
56.5 "	Florida Network	3:1	Ø μs	NB
62.2 "	Florida Network	3:1	Ø μs	NB
62.4 "	Agricultural Broadcasting Network-ABN	3:1	Ø μs	NB
63.4 "	Kansas Information Network-KIN	2:1	75 μs	NB
63.8 "	Cardinal Baseball	2:1	75 μs	NB
64.3 "	Royals Baseball	2:1	75 μs	NB
68.0 MHz	Texas State Network	2:1	75 μs	WB
71.5 "	Learfield	2:1	75 μs	NB/MCPC
72.7 "	Total Radio	2:1	75 μs	NB
74.4 "	Louisiana Network	2:1	75 μs	NB
74.7 "	Sports One	2:1	75 μs	NB
74.9 "	Sports Two	2:1	75 μs	NB
75.2 "	Mississippi Network	2:1	75 μs	NB
75.4 "	Sports Network North Carolina	2:1	75 μs	NB
77.4 "	North Carolina News Network	2:1	75 μs	NB
77.75 "	Virginia Radio	2:1	75 μs	NB
81.4 "	Arkansas Radio KARN	3:1	Ø μs	NB
83.0 "	Voice Network Texas	2:1	75 μs	NB

FREQUENCY LOCATIONS OF SCPC SERVICES (cont'd)

WESTAR 3, T2 (cont'd)

FREQUENCY	SERVICE	COMPANDING	PRE-EMPHASIS	BANDWIDTH
84.4 "	Greenbay Packers Network	2:1	75 μs	NB
85.9 "	Portland or Trailblazers	2:1	75 μs	NB
86.1 "	Milwaukee Brewers Network	2:1	75 μs	NB

WESTAR 4, T2D

FREQUENCY	SERVICE	COMPANDING	PRE-EMPHASIS	BANDWIDTH
54.0 MHz	NPR 14 Sports (AT)	3:1	Ø	NB
54.3 "	NPR 13 Sports (AT)	3:1	Ø	NB
54.7 "	NPR 15 Sports (AT)	3:1	Ø	NB
55.2 "	NPR 20 Sports (KC)	3:1	Ø	NB
55.6 "	NPR 19 Sports (EL)	3:1	Ø	NB
55.9 "	NPR 18 Sports (CI)	3:1	Ø	NB
56.3 MHz	NPR Sports (SE)	3:1	Ø	NB
56.7 "	NPR 23 Sports (SF)	3:1	Ø	NB
57.1 "	Alanet	3:1	Ø	NB
57.6 "	NSP (F5) Paging	3:1	Ø	NB
57.9 "	NET–BH	3:1	Ø	WB
62.6 "	Mutual	3:1	25 μs	WB
63.6 "	Georgia Radio News	3:1	Ø	NB
64.0 "	Mutual 5	3:1	25 μs	WB
64.4 "	Mutual 3	3:1	25 μs	WB
64.9 "	NPR 12	3:1	Ø stereo	WB
65.6 "	NPR 11	3:1	Ø stereo	WB
66.0 "	NPR 10	3:1	Ø	WB
66.4 "	Mutual	3:1	25 μs	NB
66.7 "	NPR 27	3:1	Ø	NB
67.1 "	Mutual	3:1	25 μs	NB
67.4 "	NPR 9	3:1	Ø	WB
67.7 "	NPR 8	3:1	Ø stereo	WB
68.3 "	NPR 7	3:1	Ø stereo	WB
71.1 "	NPR F2	3:1	Ø	NB
71.9 "	NPR 6	3:1	Ø	WB
72.2 "	NPR F1	DACS/NETCUE	Ø	NB
72.5 "	NPR 5	3:1	Ø	WB
72.8 "	NPR 4	3:1	Ø	WB
73.8 "	Georgia Radio Sports	3:1	Ø	NB
74.2 "	NPR 3	3:1	Ø	WB
74.6 "	NPR 2	3:1	Ø	WB

FREQUENCY LOCATIONS OF SCPC SERVICES (cont'd)

WESTAR 4, T2D (cont'd)

FREQUENCY	SERVICE	COMPANDING	PRE-EMPHASIS	BANDWIDTH
75.3 MHz	NPR 1	3:1	Ø	WB
75.8 "	Mutual 1	3:1	25 μs	WB
76.3 "	Minn. Radio News	2:1	75 μs	NB
76.7 "	Mutual 4	3:1	25 μs	WB
77.1 "	Minn. Radio News	2:1	75 μs	NB
77.9 "	Mutual 2	3:1	Ø	WB
83.5 "	NET-BH	3:1	Ø	WB
83.8 "	Alanet	3:1	Ø	NB
84.1 "	NPR 16 Sports (BO)	3:1	Ø	NB
84.4 "	NPR 22 (NY)	3:1	Ø	NB
84.7 "	NPR 25 (SP)	3:1	Ø	NB
85.1 "	NPR 17 (CH)	3:1	Ø	NB
85.6 "	NPR 26 (WA)	3:1	Ø	NB
86.1 "	NPR 21 (LA)	3:1	Ø	NB

NB-Narrow Band

WB-Wide Band

MCPC-Multiple Carrier Per Channel -FDM/SCPC RTTY CHANNELS

CHAPTER FIVE

SATELLITE NETWORKS

HISTORY OF DATA COMMUNICATIONS

The term data communications is used to describe an environment in which data is produced by a terminal or computer and is transported using communications facilities, such as phone lines, microwave, optical fibers and satellites.

The means of long distance communications has been in existence for over 100 years. The telegraph was conceived in the early 1800's, followed by the telephone in the 1870's.

In the early 1940's, the military attempted to place into use an inventory control system using a combination of data processing and communications. Later around 1952, American Airlines developed the first commercial data communications system to handle airline reservations. In today's business world, many data communications systems are used by airlines, banks, stockbrokers, insurance companies; and by many other businesses, large and small.

The new means of communication is the Satellite System. With each new development, the cost of satellite communications has dropped and its use increased.

Many of the common systems of data handling are now being transmitted via satellite at moderate cost. These are the same systems that several years ago were transmitted by telephone line, microwave communications, high frequency radio links, and transoceanic cable systems.

All data communications have common characteristics and employ the following:

(A) *ON LINE TERMINALS* that receive or transmit information and are said to be "on line." These terminals are connected and send data directly to a computer (processor) of another terminal without being stored on disk or tape. If the data is stored on a system before direct transmission, the terminal is considered "off line."

(B) *REAL-TIME PROCESSING* requires that the data be acted on immediately after receipt by the other terminal. A computer reduces the number of units in inventory immediately upon the data receipt, showing the number of units sold or delivered. Most data communications systems are used on-line and are real-time processing. Banks send data on cleared checks on an account and subtract the amount of the check at the time of receipt of the data.

(C) *FAST RESPONSE TIME*—Most data communications systems provide an immediate response to inquiries. This time is measured from the time the data is entered (enter key activated) until a reply is received. One to two seconds is considered fast, two to ten seconds is an average response; more than ten seconds is considered slow response time.

(D) *MULTIPLE SIMULTANEOUS USERS*—Most data communications systems serve many users at the same time; United Press International, Reuters Services, Dow-Jones and Airlines Reservations Services. All of these services would not be economically feasible if only one person could access and use the service at one time. These systems must be able to handle many users at the same time. **Figure 5.1.**

TELECONFERENCING VIA SATELLITE

VideoStar Connections is a satellite communications company, which provides complete communication services for satellite teleconferencing to first-class hotel meeting rooms throughout the U.S., and also to college and university sites through an agreement with the National University Teleconference Network (NUTN). The company takes total responsibility for transmission of domestic or international events, including all the necessary satellite communications facilities and the voice return circuits for two-way audio if desired. VideoStar's services include booking satellite time, arranging satellite facilities from the conference origination point, providing satellite receiving stations at the various meeting locations, and providing large screen video displays in the meeting rooms. **Figure 5.2.**

Using a combination of fixed and transportable satellite earth stations, VideoStar's Tele-Meeting Network now serves all major cities around the country. Additional cities are being added continually, and through agreements with overseas affiliates, VideoStar can route a teleconference signal to or from anywhere in the world. **Figure 5.3, Figure 5.4.**

Besides hotels, teleconferences can be delivered to some convention centers, civic auditoriums, and even directly to the client's offices. All of our facilities are staffed with experienced satellite field engineers, and include the capability to receive from any domestic satellite so that televised events have acceptable signal security, high reliability and complete scheduling flexibility. VideoStar Connections, Inc. 3390 Peachtree Rd., Atlanta, Ga., 30326.

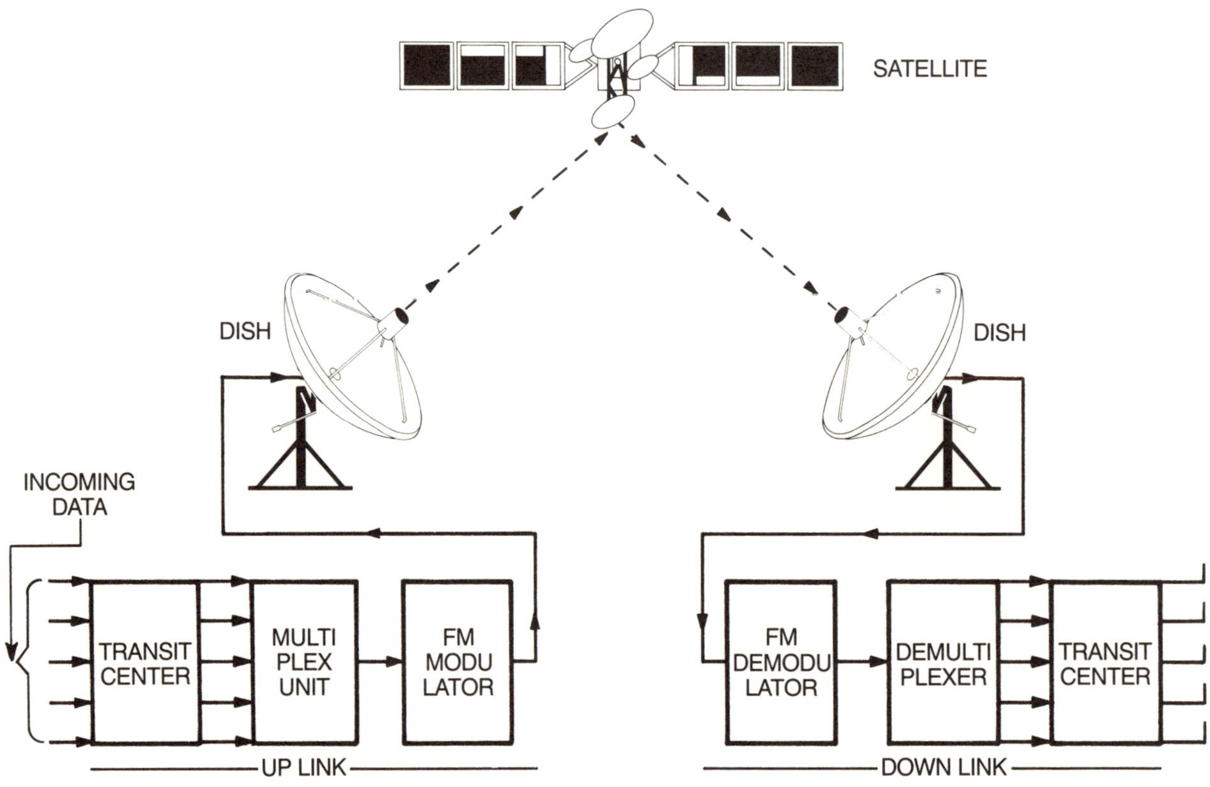

Figure 5.1 *Satellite data system using FDM/FM.*

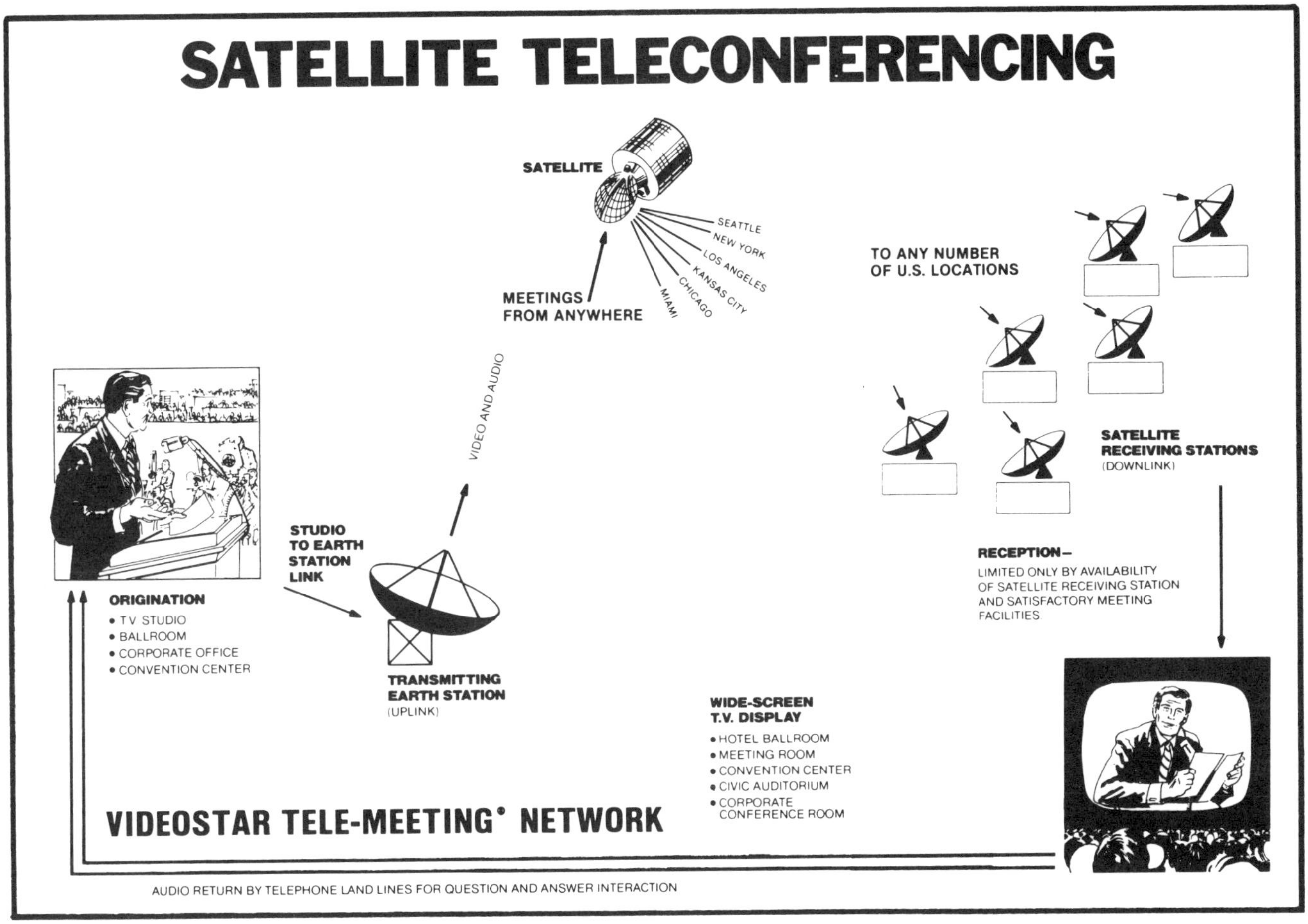
SATELLITE TELECONFERENCING
SATELLITE
SEATTLE
NEW YORK
LOS ANGELES
KANSAS CITY
CHICAGO
MIAMI
MEETINGS FROM ANYWHERE
TO ANY NUMBER OF U.S. LOCATIONS
VIDEO AND AUDIO
SATELLITE RECEIVING STATIONS (DOWNLINK)
STUDIO TO EARTH STATION LINK
ORIGINATION
• T.V. STUDIO
• BALLROOM
• CORPORATE OFFICE
• CONVENTION CENTER
TRANSMITTING EARTH STATION (UPLINK)
RECEPTION– LIMITED ONLY BY AVAILABILITY OF SATELLITE RECEIVING STATION AND SATISFACTORY MEETING FACILITIES
WIDE-SCREEN T.V. DISPLAY
• HOTEL BALLROOM
• MEETING ROOM
• CONVENTION CENTER
• CIVIC AUDITORIUM
• CORPORATE CONFERENCE ROOM
VIDEOSTAR TELE-MEETING® NETWORK
AUDIO RETURN BY TELEPHONE LAND LINES FOR QUESTION AND ANSWER INTERACTION

Figure 5.2 *Videostar tele-meeting® network.*
Courtesy Videostar, Inc.

VIDEOSTAR
THE SATELLITE NETWORKING COMPANY

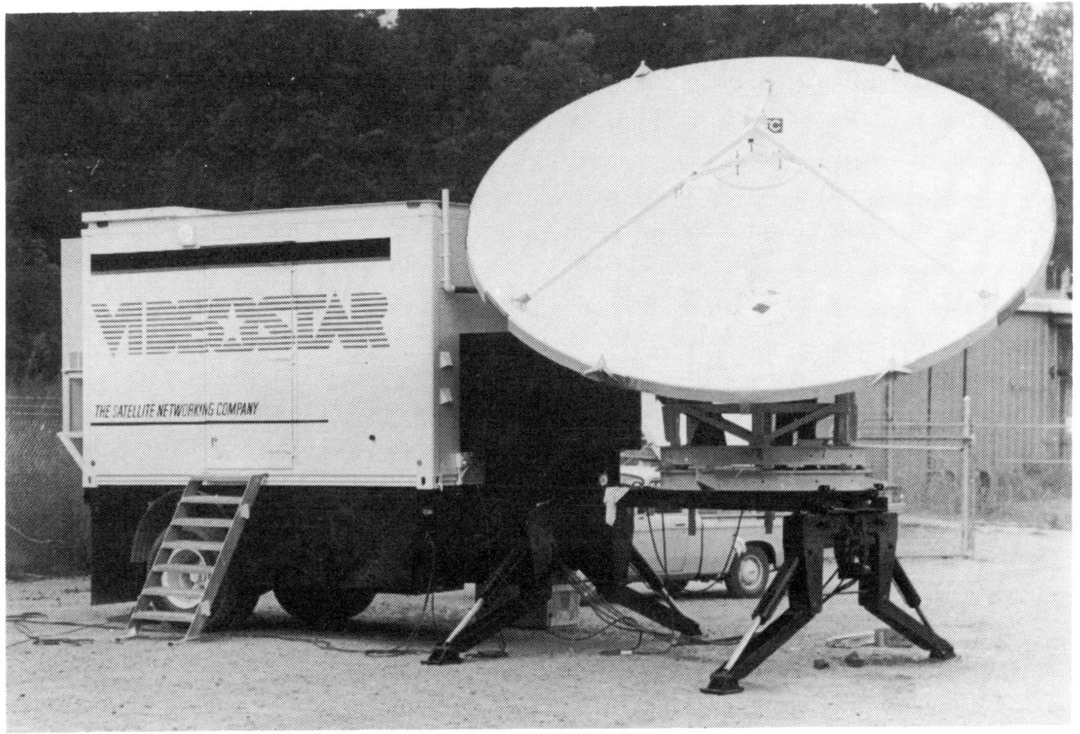

Figure 5.3 *Fully transportable uplink.*
Courtesy Videostar, Inc.

DIGITAL RADIO NETWORKS VIA SATELLITE

Today all of the major radio networks are either now using satellite to distribute their programming to their affiliates, or are in an active transition program into satellite delivery. This switch from landlines to satellite has been prompted by several factors: The increasing costs of landlines and the line quality variance between different locations. The break-up of AT&T brought on a degree of uncertainty to the networks which had the problem of serving many sections of the country. This break-up now meant that the networks were forced to deal with many distributors instead of one. A further incentive was the decreasing costs of satellite space units and decreasing costs of earth stations (downlink).

The overall trend in radio networking stressed improvement of overall quality of transmission and more flexibility for the affiliates in the future, with the possibility of supplying other types of services for added revenue for the network and the affiliate. Several large networks now plan on utilizing their satellite distribution systems to open the door to nationwide paging systems.

Other added use of these systems would be:

- Internal communications between the networks and its affiliates.
- Printout of internal data and communication from the network.
- Selective or global message service, the ability to communicate selectively to affiliates in any combination required via addressing the category of stations on a regular basis, or by need.
- The future use of the affiliates rebroadcasted subcarrier groups (SCA) for date distribution and other uses.

The Audio Digital Distribution Service (ADDS) has been in use now for radio networking since 1983 with good success. This same service (ADDS) is referred to by another system name: Digital Audio Earth Station (DAES) which is one-half of the total system. The Audio Digital Distribution Service (ADDS) is the transmitting or distribution end of the total network system. The receive end or downlink side is known as Digital Audio Earth Station (DAES) **Figure 5.5.**

The ADDS system makes use of a number of

Figure 5.4 *Interior view of transportable uplink van.*
Courtesy Videostar, Inc.

techniques: end to end digital, bit-rate companding, time division, multiplex, pulse code modulation techniques, Forward Error Correction (FEC), plus other methods.

After several years of study, it was decided to use a combination of these advanced techniques for transmitting the needed high-quality radio network audio. At this point, the Scientific-Atlanta Company started a program to develop specific equipment for this use.

The networks, ABC, CBS, NBC and others are recommending the direct purchase of the needed ADDS equipment from Scientific-Atlanta Company. The network furnishes and pays for the satellite time, and the affiliates purchase their needed downlink (receive) equipment.

In the case of the RKO radio network, the network will pay for and install the needed equipment for its stations located in the 150 largest markets.

The equipment that the stations buy and install will be the basic S/A equipment with a few options available, depending on what each affiliate station's needs are, and largely depending on its programming plans.

The table below lists the perimeters for the four networks.

DIGITAL SATELLITE NETWORK EQUIPMENT (ADDS)

The Scientific-Atlanta Company designates its equipment package for ADDS reception a digital audio earth station (DAES).

This configuration consists of S/A series 9000 antenna, 2.8 meter, a series 300–1, 120 degree Kelvin low-noise amplifier (LNA), 200 feet of ½ inch foam-filled coax cable, the model 7300 wideband BPSK receiver, and the model 7325 digital processing unit (DPU). This unit will house up to 7 cards of any combination for the desired services. A full description of the S/A DAT-32 digital audio terminal follows.

The Scientific-Atlanta DAT-32 Digital Audio Terminal is designed for reception of media information broadcast via satellite using digital data techniques. With this terminal, it is possible to receive various types of data services including:

- Digital Program Audio Channels (15 kHz or 7.5 kHz)
- Voice Cue Channel (32 kb/s)
- Data Channels (32 kb/s)

The DAT-32 is designed to receive a single biphase shift keying (PBSK) carrier modulated by a high-speed Time Division Multiplex (TDM) digital data. It demodulates and demultiplexes the data into the audio

	ABC	NBC	CBS	RKO
Transponder No.	23	19	19	19
Transponder Freq.	4160 MHz	4080 MHz	4080 MHz	4080 MHz
No. of Network Channels	19	6	6	8
Sug. Plug-in-Channel Units	2 Dual 15 KHz Audio Units 1 Voice Cue Unit	1 Dual 15 KHz Audio Unit 1 Dual 7.5 KHz Audio Unit 1 Voice Cue Unit 1 Data Unit	1 Dual 15 KHz Audio Unit 1 Dual 7.5 KHz Audio Unit 1 Voice Cue Unit 1 Data Unit	2 Dual 15 KHz Audio Unit 1 Voice Cue Unit

and data channels. The system achieves a bit error rate (BER) of less than 10^{-7} with over 1.5 dB of margin for most locations throughout the continental United States.

The data is distributed using a BPSK modulation at 8.78 Mb/s rate. This rate and modulation format were selected especially to provide sufficient spread of transmitted energy to be cleared by FCC for transmission, yet narrow enough to fit snugly between two terrestrial TD–2 potential interfering signals located at ± 10 MHz from the transponder center frequency. TDM signal carrier digital transmission allows saturated transponder operation, so all available power is used to provide minimum network annual cost for point-to-multipoint, one-way distribution into 2.8 meter terminals with a large variety of data services. The satellite transponder has the capacity to provide 20 each 15 kHz (or 40 each 7.5 kHz) channels each of which can be divided into 12 subchannels for voice cue and data uses.

The standard DAT–32 receive-only earth terminal includes the Series 9000 2.8 meter parabolic reflector, a Model 300–1 120 degree Kelvin low-noise amplifier (LNA), 200 feet of 0.5 inch foam-filled coaxial RF cable, the Model 7300 wideband BPSK receiver and the Model 7325 digital processing unit (DPU), which houses any combination of up to seven program audio, voice cue and data channel cards for the desired services.

Both the 7.5 kHz and the 15 kHz program audio modules are dual channel cards. A separate card is required for voice cue. Each data card provides for three output channels of asynchronous information. The LNA is mounted at the antenna focal point, while the two 10.5 inch electronics chassis are installed into a standard 19.5 inch equipment rack in the host facility. An extension chassis is also available when additional channel units are required. DC power to the LNA is provided through the coaxial cable from the receiver.

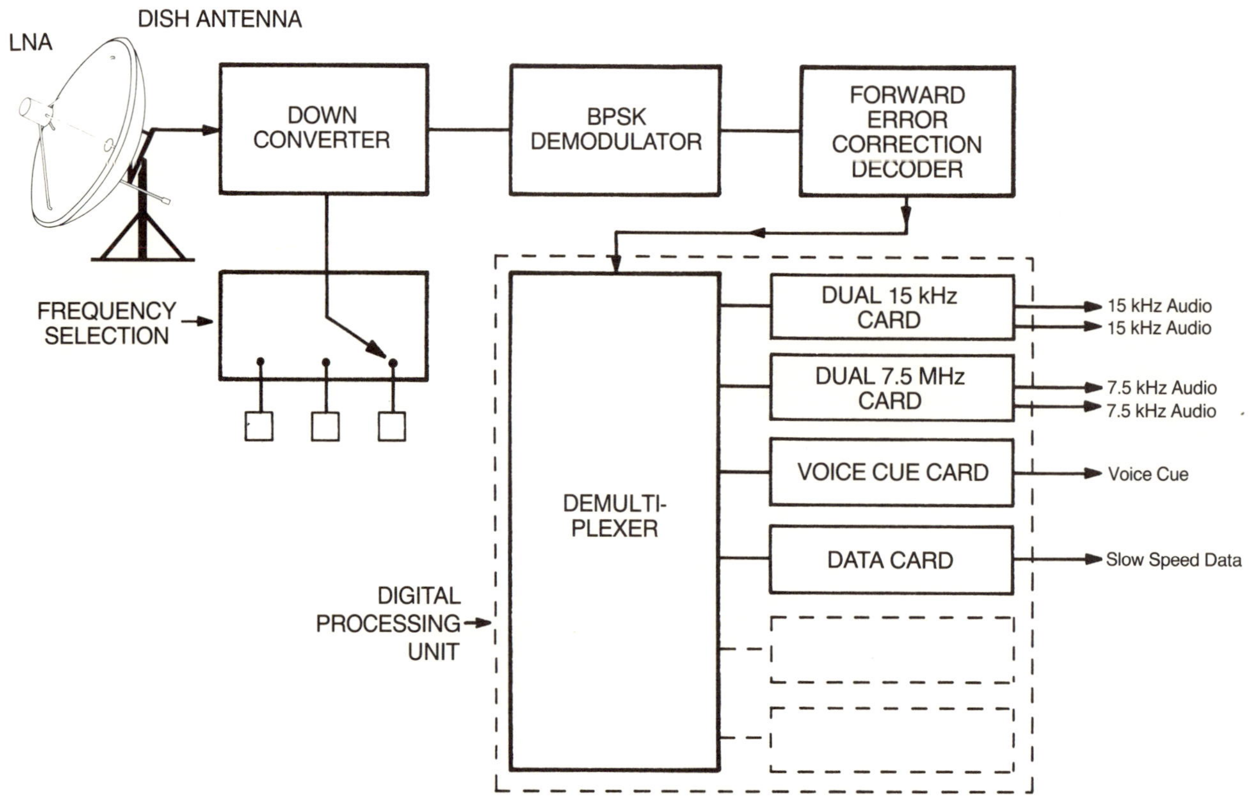

Figure 5.5 *Digital audio earth station setup (radio station).*

The downlink signal is received by the antenna. The RF energy is reflected to the LNA which is mounted at the antenna focal point. This device provides the first stage of amplification so adequate signal is imput to the receiver through the coaxial cable. The BPSK receiver-decoder downconverts the 4 GHz signal to a 70 MHz intermediate frequency, demodulates the wideband IF BPSK signal, and provides Forward Error Correction Coding at a ⅞ rate. The resultant 7.68 Mb/s data stream is passed to the DPU chassis. Frame sync and descrambling occur in the demultiplexer card provided in this unit. The individual data streams are routed to the channel cards, where reconstruction of the analog signal takes place. This signal is then amplified for interconnection to the appropriate radio station equipment.

The 15 kHz program signal is sampled at a 32 kHz rate (16 kHz for a 7.5 kHz signal), and digitized into a 15–bit word. Digital companding techniques are used to compress instantaneously the 15–bit word into an 11–bit word to reduce the transponder bandwidth required. A parity bit is included resulting in a word length of 12 bits.

The parity bit is used in an additional error concealment encoding that allows the bit-error rate to degrade to 10^{-5} before errors are "just preceptible" in the 15 kHz program audio output. Each program audio card contains two separate 15 kHz (or 7.5 kHz) channels which may be used independently or as a stereo pair. The balanced output level is 24 dBm maximum into a 600 chm load. This allows for 16 dBm of headroom when a 8 dBm average program level (APL) is used. Maximum deviation from a flat frequency response is 1.0 dB over the 40 Hz to 15 kHz (or 7.5 kHz) band. The unit provides in excess of 80 dB peak signal-to-idle noise ratio and a peak signal-to-quantization noise ratio of 56 dB.

The voice cue channel is used for network closed circuit broadcasts and event coordination. A continuous variable slope delta-mod (CVSD) encoding scheme is employed for voice quality service at a 32 kb/s rate.

The data decoder will simultaneously process three asynchronous RS–232 data streams using a single 32 kb/s subchannel. This unit is fully addressable from the originating studio. The data channel can be used for electronic mail and other distribution of one-way data. **Figure 5.6** and **Figure 5.7.**

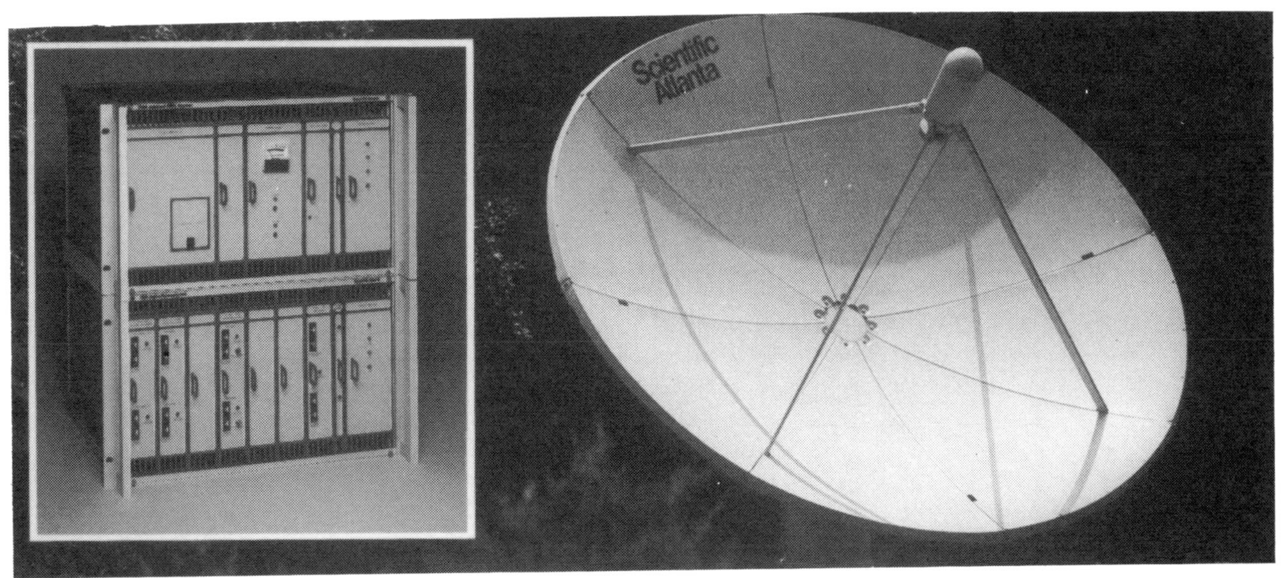

Figure 5.6 *Scientific–Atlanta's DAT–32 digital audio terminal.*
Courtesy Scientific-Atlanta, Inc.

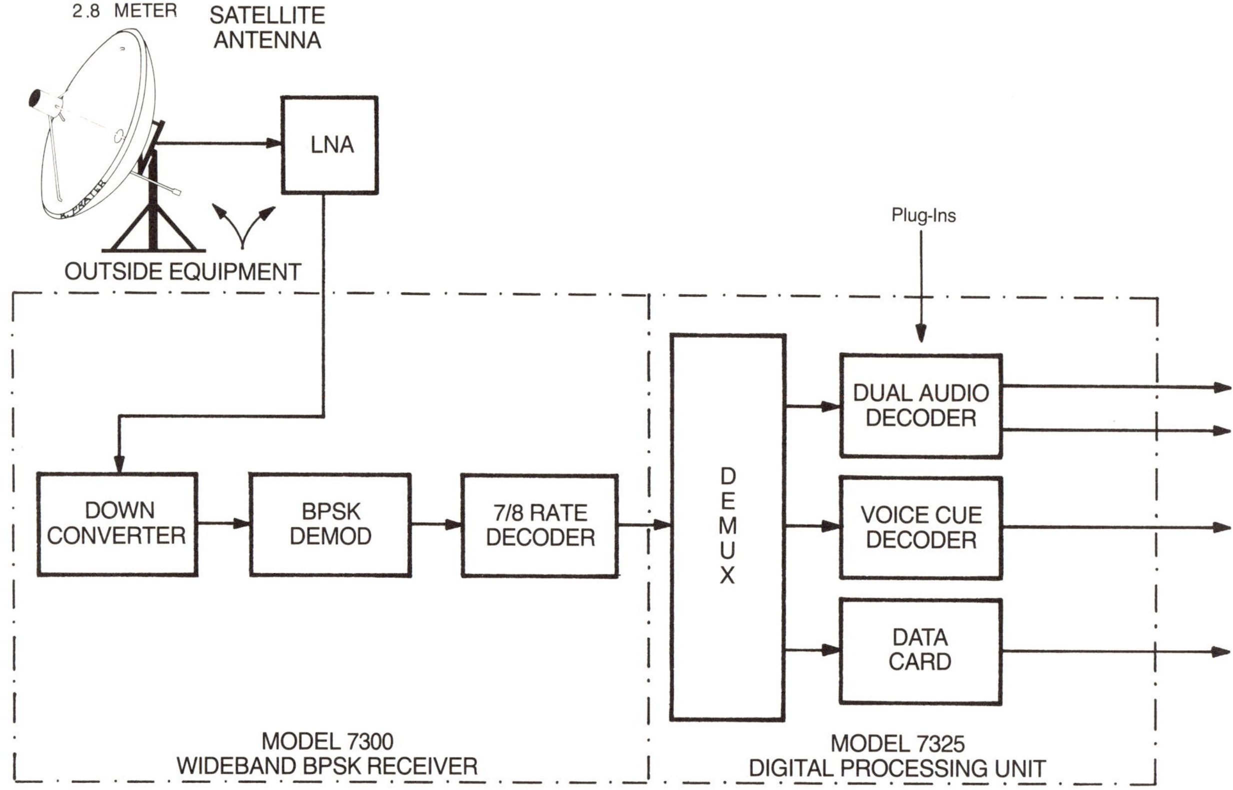

Figure 5.7 *Scientific-Atlanta DAT-32 system block diagram.*

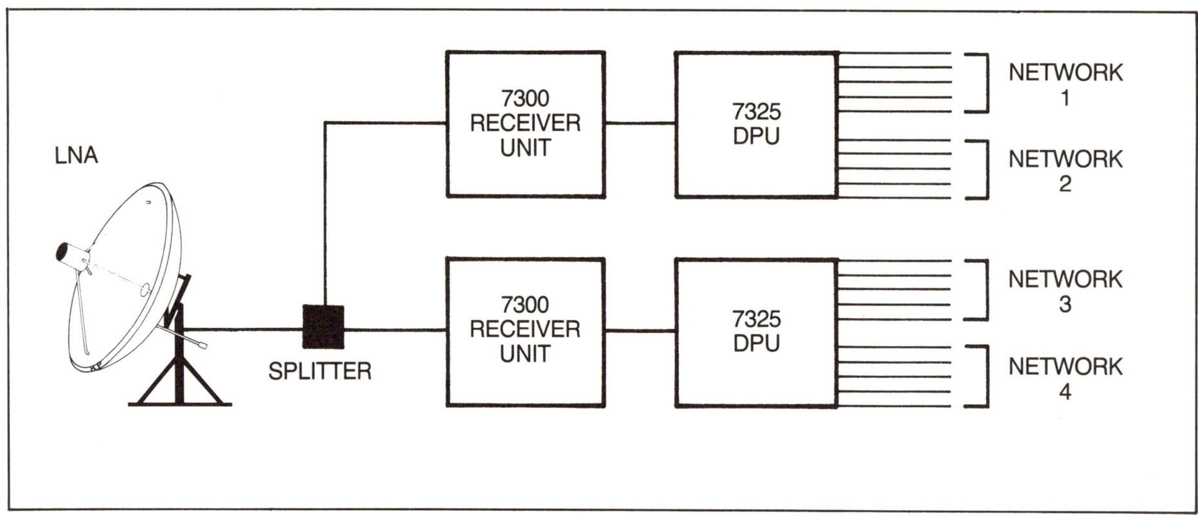

Figure 5.8 *Dual electronic system.*

A redundant DAT–32 system can be had by adding another 7300 receiver to the system with its own digital processing unit, as shown in **Figure 5.8.** Here two 7300 receivers are fed in parallel from the LNA, and in turn each receiver feeding its own DPU. This makes a very flexible system. For example — if one receiver is dedicated to transponder 23 reception, and the other receiver dedicated to transponder 19 reception, it is then possible to bring up all of the 20 audio outputs at one time for whatever use they want to be put to. Of course, this system would have to be placed on say ABC networks and the others would have to be RKO, CBS or NBC. **Figure 5.9.**

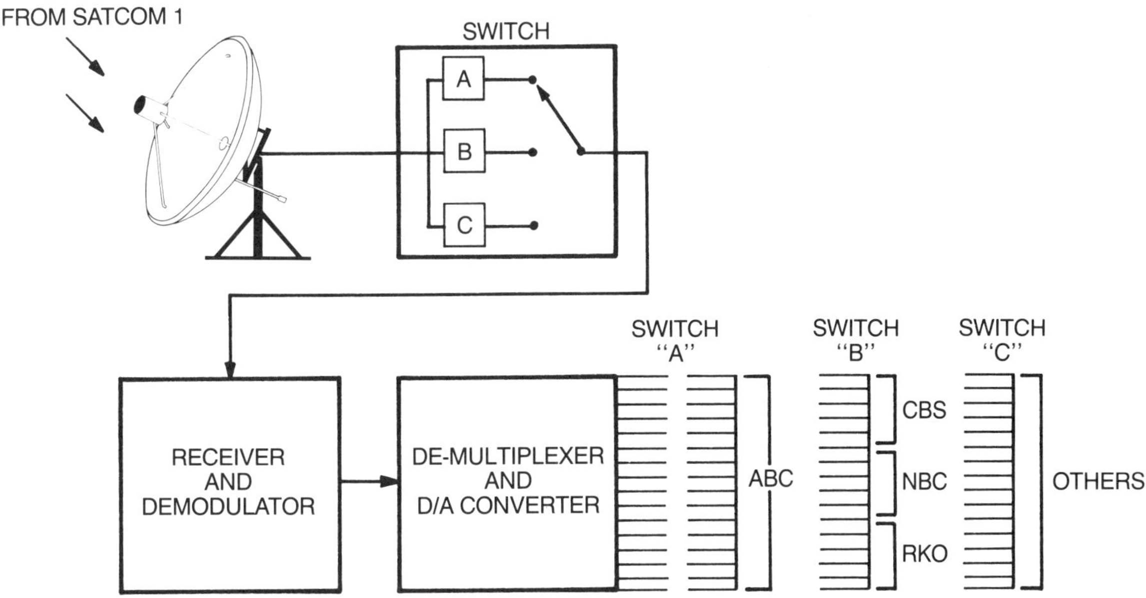

Figure 5.9 *DAES with optional transponder switch to receive all the channels of all four networks.*

MODULATION ASSOCIATES CORPORATE DATA NETWORK

The SU–10 SOLID STATE UPLINK is specifically designed for corporate data networks, regional radio networks, remote broadcasting, and data collection — as well as for telephone and compressed video transmission. For data networks the uplink allows full duplex or point-to-multipoint high speed transmission of data information. In radio networks the SU–10 is ideal for two independent SCPC uplink channels or for stereo transmission.

The SU–10 design incorporates 10 Watt solid state, high power amplifiers. The amplifier outputs are combined into a single waveguide output for application to the uplink antenna. In its complete configuration the SU–10 is available with audio or data processors, frequency agile modulators, dual channel upconverter, and dual HPA's. The entire uplink is housed in a 5–foot rack for easy shelter mounting right at the antenna.

The low phase noise and high frequency stability of the upconverter LO source allows transmission of SCPC program or data channels over satellite links without the need for pilot channels at the downlink. Image reject mixers and output filtering keep spurious and unwanted signals well below allowable FCC limits.

Optional features include redundancy and dial up remote control from your personal computer — using the MODULATION ASSOCIATES UC–10 Remote Controller. Up to 16 agile modulators can be switched off and on, and moved to new frequency assignments. Alarm limits for RF power, voltage, temperature, audio level, and loss of data can be set and cause the UC–10 to dial in alarms to the master control. A 4-to-1 remote control audio switch is available for unmanned remote broadcast uplinks. The user can interface other types of baseband equipment at the IF input of the converter.

A truly portable version of the SU–10 is available for remote broadcasts, compressed color video teleconferences, data collection, and temporary telephone service. This SU–10 portable is housed in a 4–foot by 2–foot by 2–foot roll-around, carrying case which is accepted as luggage or cargo by most airlines. **Figure 5.10.**

The M/A Satellite Data Receiver was designed specifically for use in corporate data, electronic mail, and reservations networks — the MODULATION ASSOCIATES DATA-SAT provides convenient, low-cost, and reliable satellite reception of high volume, point-to-multipoint 56 kbs data information, permitting universal high quality service to any geographic location within a satellite coverage area at a rate five to eight times faster than the average landline rate and with fewer errors.

The DATA-SAT is designed for applications where a large number of highly reliable data downlinks are required. The DATA-SAT takes in a satellite signal at 4 GHz, downconverts it to 70 MHz, and demodulates the data stream. It uses commonly available components, and is designed to operate for years without on site maintenance support. A front panel, LED display indicates whether the data is good, marginal, or unusable. A sophisticated AFC tracks out any frequency errors in transmission.

Because the DATA-SAT is a stand-alone, complete receiver, data capability can be added to a user's already existing earth terminal at any time without complex and expensive common equipment. **Figure 5.11.**

With the M/A SU–10 Uplink and DR–56 units, any number downlink systems can be in place to serve most corporate needs, large and small. This system could serve companies countrywide, that have many branch offices and warehouses. **Figure 5.12.** These network experts will assist the company with a complete satellite package, including, space segment, FCC licensing, hardware and installation of the complete system, providing low-cost, reliable data reception to multiple users.

SPREAD SPECTRUM TECHNOLOGY

The cost and size of earth stations have, in the past, prohibited small users from taking advantage of satellite communications. As recently as 10 years ago, an earth station 32 feet in diameter was considered "small." Currently, the term "small earth station" generally means one approximately 16 feet in diameter. However, the latest size reduction produced a disk measuring two feet in diameter and weighing 15 pounds, facilitating installation and maintenance.

The low-cost receiving earth station, located at the user's home or office, eliminates the need for both local and long-distance telephone charges. Selective data filtering by the earth station permits relevant portions of larger databases to be retained for local searching and display. The use of microcomputer intelligence at the user site reduces large-computer time-sharing complexities and permits immediate

Figure 5.10 *Modulation Associates SU-10 solid state uplink unit.*
Courtesy Modulation Associates, Inc.

Figure 5.11 *Modulation Associates data SAT receiver.*
Courtesy Modulation Associates, Inc.

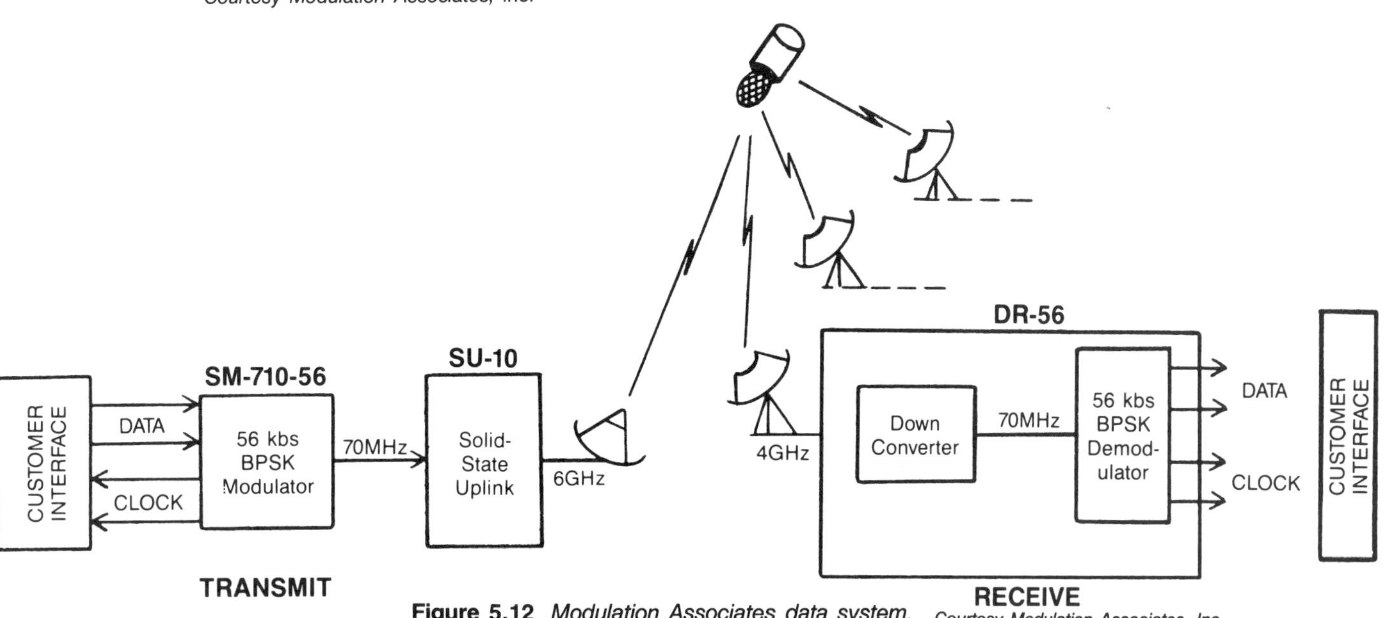

Figure 5.12 *Modulation Associates data system.* *Courtesy Modulation Associates, Inc.*

nationwide local access to centrally updated data files without shared-computer response-time delays.

Packet-switching point-to-multipoint satellite networks are available to distribute databases to microcomputers (or minicomputers) at a large number of widely-dispersed remote sites, while retaining central database control.

Spread spectrum is a transmission modulation technique that has long been used in radio astronomy, research and military communications applications. The technique has been in regular use in commercial data communications networks only since 1981.

Spread spectrum is particularly suitable for communications via satellite to small earth stations because it permits reliable data transmission despite the presence of strong interfering signals. This technique has been used in military applications to transmit unobtrusive signals that ensure reliable transmission despite "jamming" signals deliberately transmitted to create interference. Radio astronomy and other space research applications of spread spectrum have permitted reliable communications over interplanetary distances with very weak signals that would otherwise have been overwhelmed by random noise.

As the name implies, more frequency bandwidth is required for spread-spectrum signals than for conventional narrowband signals transmitting comparable data rates. Consequently, the technique has not been extensively used in commercial terrestrial communications applications, where conservation of bandwidth is an important network criterion. In satellite communications involving small earth satellite stations, networks are limited by the availability of satellite power, rather than by availability of bandwidth. Therefore, no additional costs result from spreading the signal over a greater bandwidth that would otherwise go unused.

In spread-spectrum data transmission, each bit is subdivided into a number of pieces called "chips." Each chip is transmitted through a conventional modulation technique as if it were a bit. For example, each phase of a binary-phase-shift-keyed signal would represent a chip in spread-spectrum transmission, whereas it would represent a bit in conventional narrowband transmission. The resulting higher occurrence of chips (and corresponding phase shifts) per bit of information spreads the signal over a wider frequency bandwidth.

The chips of a spread-spectrum transmission are ordered in a unique code sequence, called a pseudorandom noise (PN) sequence. Each spread-spectrum receiver includes a "match filter" (matched to the transmitting PN sequence). The result is that even when a large number of chips are garbled in transmission, the receiver can still perform the pattern recognition analysis to reliably recognize the bits.

The spread-spectrum technique makes possible very-small-earth-station data reception. The smaller the diameter of the receiving antenna, the wider the angle of view of that earth station. Larger earth stations, like larger telescopes, permit a narrower focus. A two-foot-diameter earth station at satellite radio frequencies of 4 GHz has a half-power beam width of 9 degrees when pointed at the equatorial orbital arc where the satellites are located.

Therefore, with satellites spaced 4 degrees apart, as they currently are, or 2 degrees apart, as the FCC has recently proposed, each small earth station is susceptible to interference from multiple satellites as well as from terrestrial microwave. The interference rejection property of spread-spectrum transmission permits small earth stations to coexist with terrestrial microwave in urban environments and permits them to operate not only with the current satellite spacing, but also with proposed narrower satellite spacings.

INTERACTIVE DATA COMMUNICATIONS NETWORK SERVICES BY EQUATORIAL COMMUNICATIONS COMPANY

Equatorial Communications Company markets satellite network products and services for terminal-to-computer, terminal-to-terminal and distributed database applications. These networks can provide end-to-end communications without the need for terrestrial circuits. The company manufactures interactive Micro Earth Stations for installation at user terminal locations as part of its satellite networks. Customer data communication services are available through EQUATORIAL's satellite transponders and value-added packet-switched satellite network. EQUATORIAL also provides turn-key private satellite data communication networks with technology licensing for large corporate customers.

EQUATORIAL's satellite networks offer independence from local and long distance telephone companies by providing a cost-effective alternative to the use of terrestrial circuits for data communications. Communication requirements of data collection, inquiry/response and database distribution applications can be effectively supported. A typical EQUATORIAL network consists of:

Transmit/receive Micro Earth Stations featuring very small antennas, and low power requirements for use with remote user data terminals.

A satellite and Master Earth Station communication link, which relays the data between the user's terminal and host computer.

A conventional high speed terrestrial or satellite communication link, connecting the Master Earth Station to the host computer.

The satellite network operates at the same speed, and with improved performance characteristics for data communications typically carried over voice grade telephone lines. EQUATORIAL offers variable inbound and outbound transponder capacity for its interactive communications service, allowing the customer to optimize the performance of his system without the need to pay for excess capacity in either direction.

Network advantages include lower cost, low data error rate, insensitivity to distance, network flexibility, simple Micro Earth Station installation, and improved network reliability. Satellite communication networks for single remote data terminals are practical. In addition, EQUATORIAL's satellite network offers extensive system diagnostics, ease of network cut-over, and network performance prediction and control.

EQUATORIAL has been providing real-time continuous transmission services for news, financial, and commodity information providers since mid-1981. Over four thousand C–100 receive-only Micro Earth Stations operating in the network demonstrate EQUATORIAL's ability to provide high quality communication services with consistent performance and availability. This prior network and earth station experience has resulted in product and service reliability enhancements that can only come after extensive field experience. EQUATORIAL has maintained availability of better than 99.9% on its common carrier spread spectrum network since start of operational services.

EQUATORIAL'S satellite network offers:

- Priority access (7 days/week, 24 hours/day) available for networks in capacity increments of 1200 bps up to 150 Kbps.
- Maximum aggregate data rate to customer interface up to 19.2 Kbps.
- Maximum aggregate data rate from customer interface up to 1200 bps.
- Bit error rate of better than 1×10^{-7}.
- Fully redundant Master Earth Station transmitter and receiver electronics, and uninterruptible power source.
- Continuous network monitoring and diagnostic analysis contributing to high network reliability.
- Proprietary spread spectrum transmission techniques maximizing noise rejection.
- Time-shared packet data communications permitting efficient satellite channel utilization.
- Multiple data channels of differing data speeds, which may be multiplexed within each customer network.
- Host computer-to-Master Earth Station high speed conventional data communication links.

EQUATORIAL's extremely high data integrity packet switched networks utilize efficient, variable-length packets. Each data packet may have either an individual or group address. Forward error correction coding is included within each packet. Sequential packet numbering combined with store-and-forward packet techniques permit retransmission of any locally uncorrectable packets.

Satellite data communications networks are available in various configurations including:

Terminal-to-host computer networks with the host computer connected to the Master Earth Station by conventional high speed satellite data link.

Terminal-to-host computer networks with the host computer connected to the Master Earth Station by conventional high speed terrestrial link.

Terminal-to-terminal interactive data communications through the Master Earth Station.

Figure 5.13 shows a terminal-to-host computer network with the host computer connected to the Master Earth Station by conventional high speed satellite link, and then to the C–200 Micro Earth Stations and associated data terminals by a spread spectrum signal from the Master Earth Station.

Figure 5.14 shows a terminal-to-host computer network with the host computer connected to the Master Earth Station by a terrestrial line. Network operation is identical to that shown in **Figure 5.10** except that the high speed satellite link is replaced by the terrestrial line.

Figure 5.15 shows a terminal-to-terminal satellite network link between two C–200 Micro Earth Stations, utilizing the Master Earth Station as an intermediate

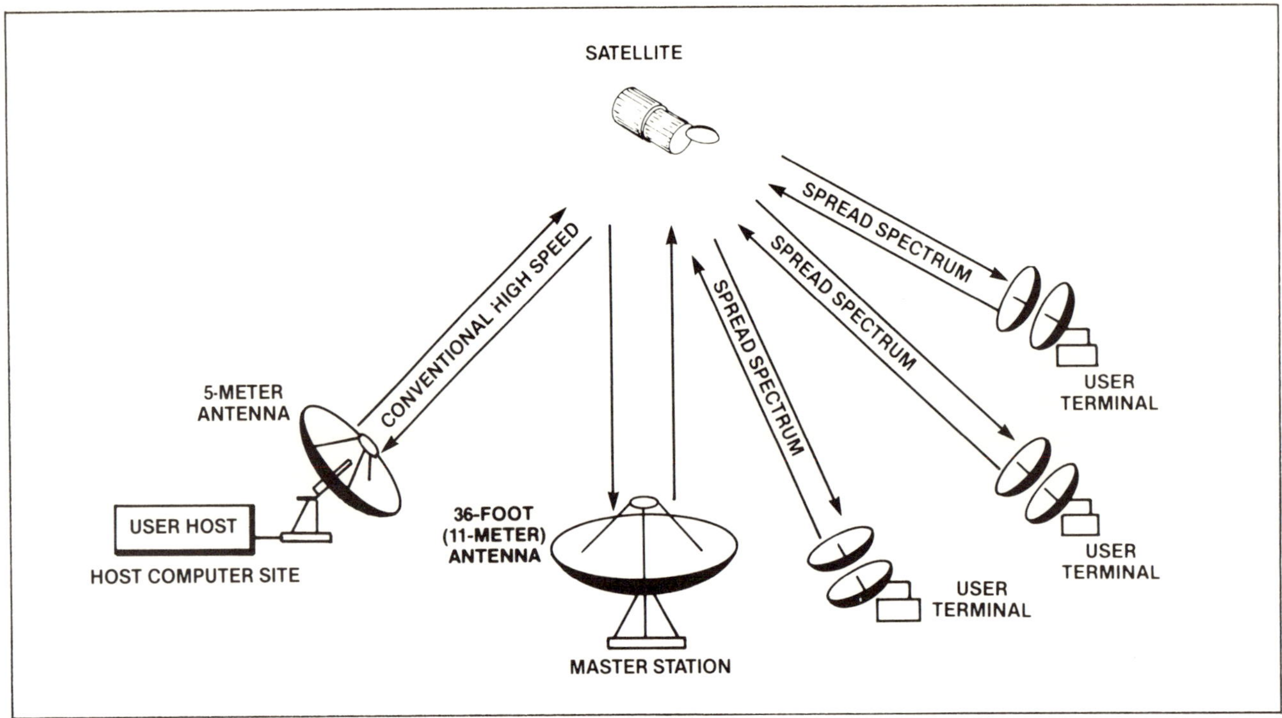

Figure 5.13 *Terminal-to-Host computer data communication network using spread spectrum and conventional satellite links.*

Courtesy Equatorial Communications, Co.

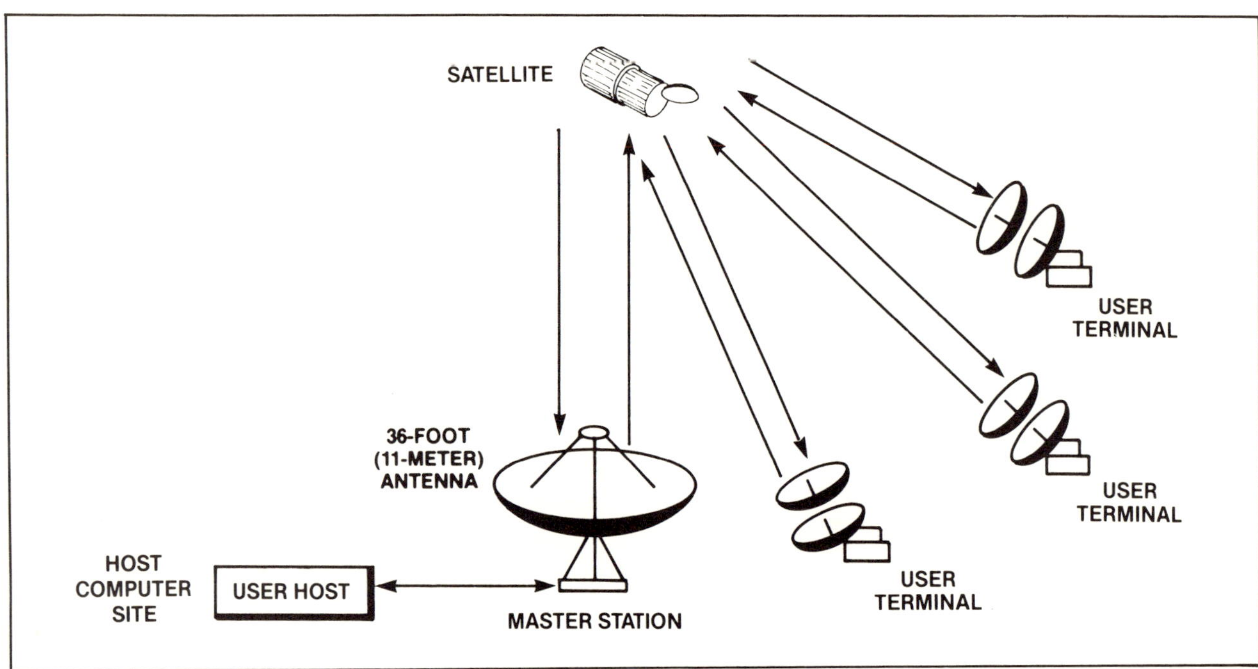

Figure 5.14 *Terminal-to-Host computer data communication network using spread spectrum satellite link and local terrestrial line.*

Courtesy Equatorial Communications, Co.

signal amplifier. The low-power signal from the first Micro Earth Station is transmitted to the large Master Earth Station, where it is amplified and transmitted back to the satellite. This strong signal is again amplified by the satellite transponder, and transmitted to the receiving Micro Earth Station with enough power to be detected by the 4–foot (1.2–meter) antenna. **Figure 5.16** — EQUATORIAL offers satellite data communication services through its Master Earth Station in Mountain View, California using transponders on the WESTAR IV satellite. This Master Earth Station, owned and operated by EQUATORIAL as part of its satellite network operations, consists of a 36–foot (11–meter) antenna, transmitting electronics, and modular arrays of switching and multiplexing electronics as shown in **Figure 5.17** and **Figure 5.18.**

Refer to the WESTAR IV footprint map **(Figure 5.19)** for contours that represent the boundaries within which the EQUATORIAL Earth Stations will operate within an error rate of 1×10^5 or better, with a 3 dB nominal margin to guard against mispointing, rain attenuation, sun noise or equipment degradation. Outside these boundaries, the earth station may receive the satellite signal and appear to operate well but the error rate will be higher. The contours have been generated from spacecraft antenna gain contours supplied by Western Union for 3840 MHz.

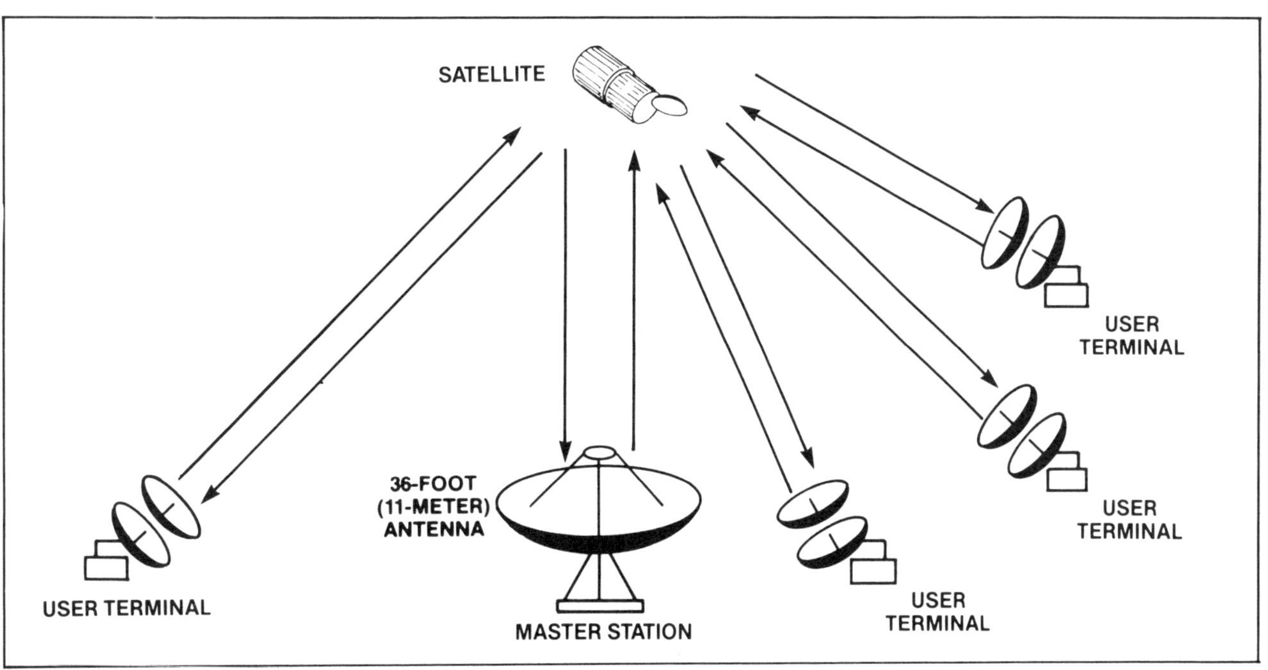

Figure 5.15 *Terminal-to-Terminal data communication network.*

Courtesy Equatorial Communications, Co.

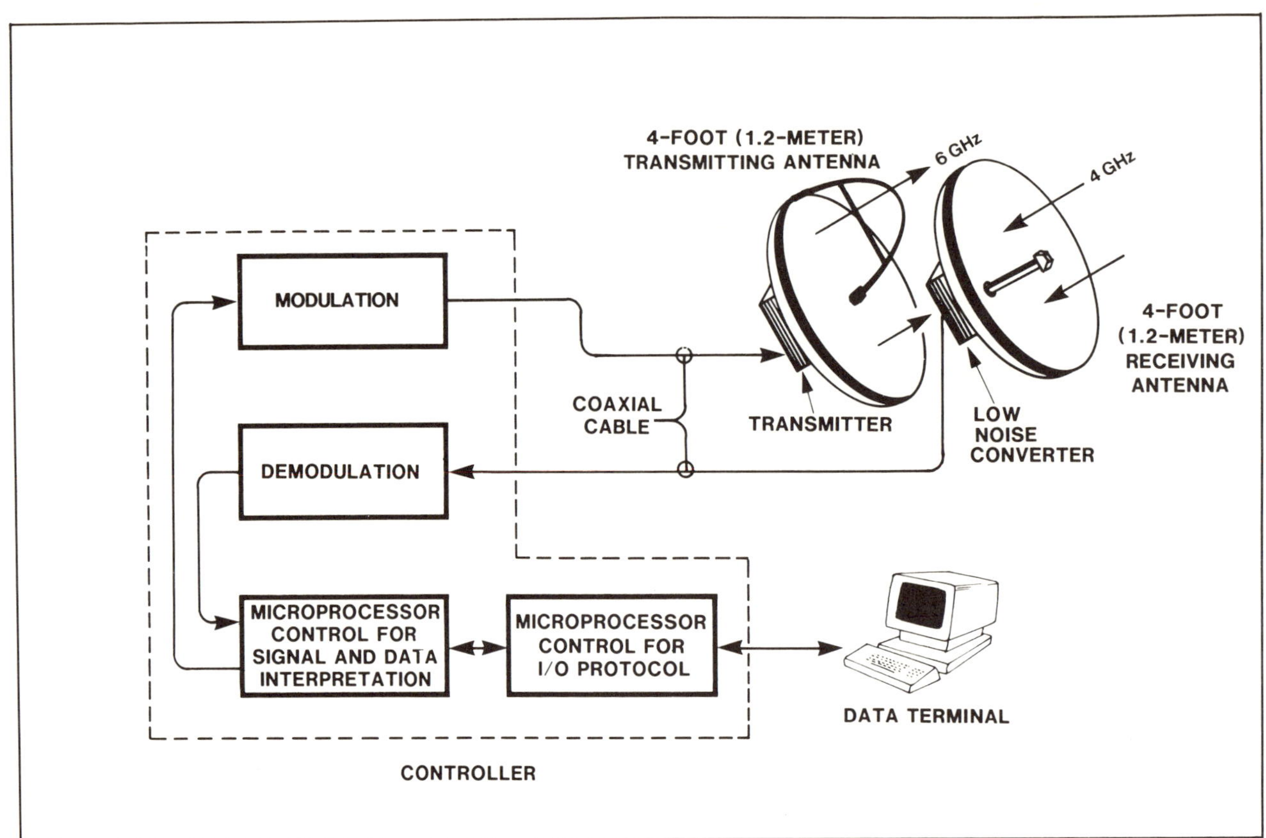

Figure 5.16 *Micro earth station block diagram.*

Courtesy Equatorial Communications, Co.

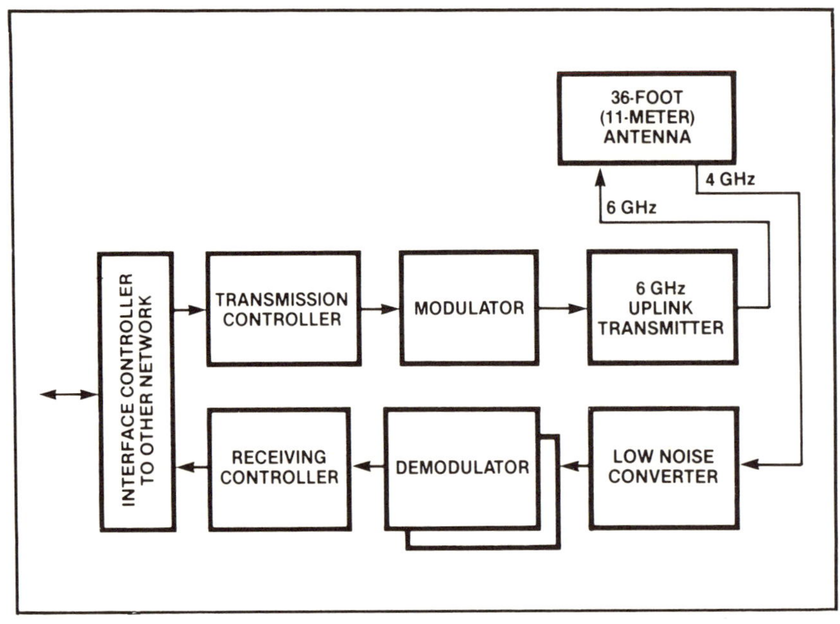

Figure 5.17 *Master station block diagram.*

Courtesy Equatorial Communications, Co.

Figure 5.18 *36 foot master earth station antenna.*
Courtesy Equatorial Communications, Co.

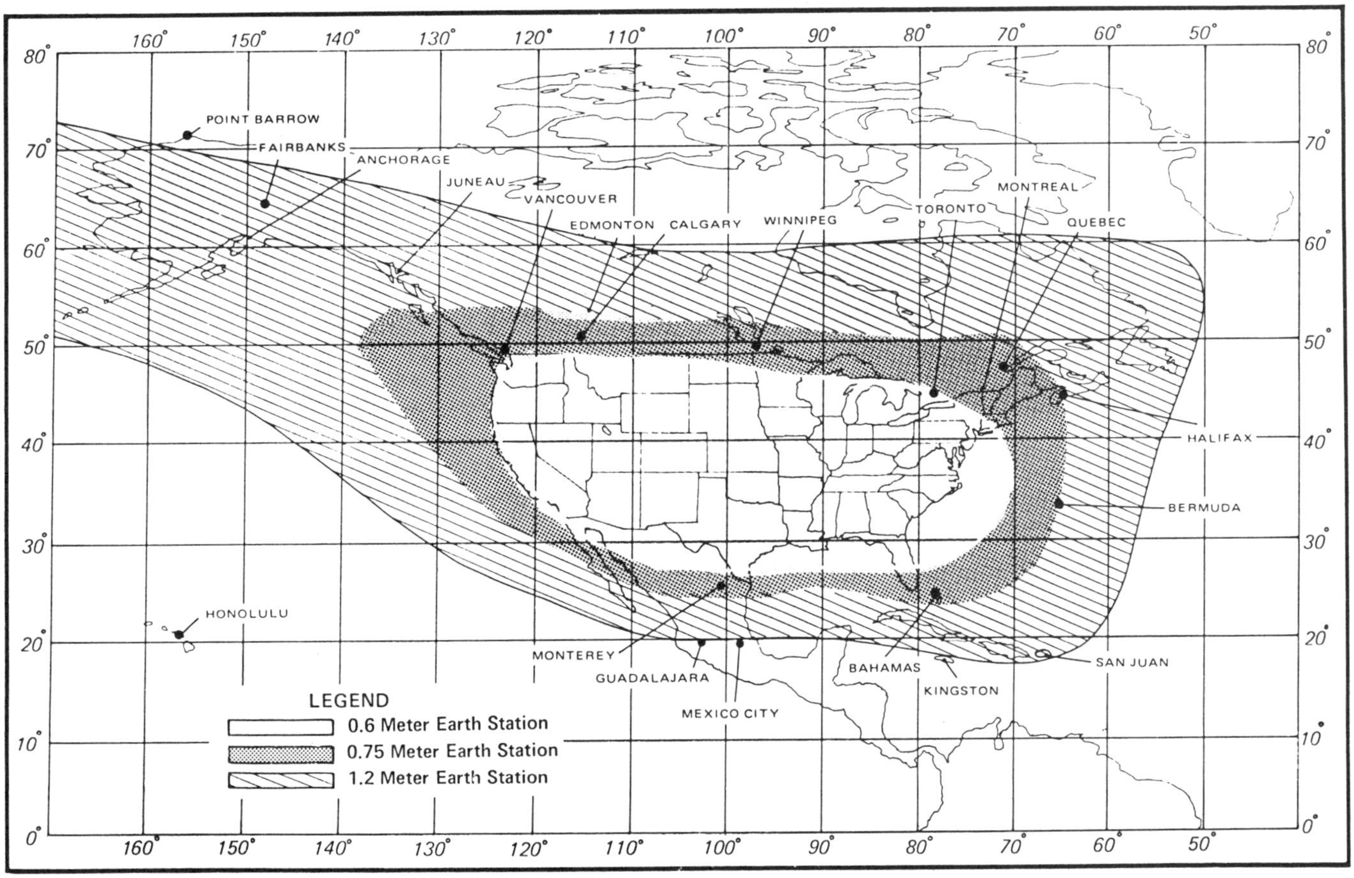

Fig. 5.19 *Westar IV Footprint*
Courtesy Equatorial Communications, Co.

EQUATORIAL's point-to-multipoint communications networks are available in various configurations including:

Point-to-multipoint networks with the data source connected to the Master Station by conventional satellite data link.

Point-to-multipoint networks with the data source connected to the Master Station by conventional terrestrial link. **Figure 5.20** shows a network with a data source connected to the Master Station by conventional terrestrial line or satellite link. The Micro Earth Stations **(Figure 5.13)** and associated terminals receive this data by a spread spectrum signal from the Master Station.

The Micro Earth Station features include:

Two-foot (0.6–meter) diameter antenna.

Lightweight.

Bit error rate better than 1×10^{-7}.

Data rates from 45 to 19,200 bits per second.

Multiport receiver capability.

Built-in intelligence/diagnostics.

Small antenna size, low power requirements and interference-resistant receiver for simple installation at urban user sites.

Reliability for continuous unattended service on end user premises.

A microprocessor provides the data handling and protocol functions necessary to adapt the Micro Earth Station to many different application requirements.

A block diagram of the C–100 Micro Earth Station is shown in **Figure 5.21.** The 2-foot (0.6-meter) antenna collects the 4 GHz signals received from the satellite. Microwave electronics in the antenna module accomplish the amplification and frequency conversion functions for the incoming signals. The separate Micro Earth Station controller module contains circuits for signal demodulation, a microprocessor based signal and data interpretation function, and 1/0 control facilities.

EQUATORIAL's high data integrity packet-switched networks utilize efficient, variable-length packets. Each data packet may have either an individual or group address. Forward error correction coding is included within each packet.

Each Micro Earth Station is identified with a unique subscriber code permitting the unit to be preprogrammed to capture selected portions of a data stream. The unit's data access authorization is maintained in non-volatile memory. Since the data authorization comes from the network manager and is transmitted over the satellite link to the Micro Earth Station receiver, the integrity of the authorization is assured. Users at the Micro Earth Station end of the system do not have access to the authorization codes.

This unique addressing capability of the Micro Earth Station provides for a flexible network control scheme. Data publishers using the EQUATORIAL satellite network may determine which Micro Earth Stations are eligible for reception of different data segments. Micro Earth Station eligibility for data reception can be changed in real-time by the network manager. Using these subscriber codes, data base publishers can deliver subsets of the data base to authorized subscribers, since each site will receive only the subsets ordered. Additional levels of coding can customize reception of individualized data or specific pages of information.

EQUATORIAL's C–200 Micro Earth Stations make satellite communication networks for single remotely located data terminals practical for the first time. These products offer significant communications costs savings in data collection, inquiry/response and data base distribution systems. Typical applications include:

- Financial Transactions
- Order Entry
- Point-of-Sale
- Process Control/Monitoring
- Reservation and Other Transportation Related Systems

EQUATORIAL's C–100 Series Micro Earth Stations make satellite communication networks for single, remotely located data terminals practical for the first time. These products offer significant communication cost savings in data distribution systems. Typical applications include:

- Financial information distribution
- Commodity news and financial information distribution
- General news delivery
- Database distribution and updating
- Electronic mail
- Digital facsimile transmission

EQUATORIAL is presently delivering Micro Earth Stations to several customers, including Reuters; Commodity News Services, Inc.; Market Information, Inc.; and Trans-Lux Corporation. Both Commodity News Services and Market Information transmit commodity news and prices to newspapers, brokers, and farmers. **Figure 5.21.**

FSK DATA ON SUBCARRIERS (SbC)

Frequency shift keying (FSK) is the system used in slow speed (low baud rates) teleprinter data networks. Some call these services radioteletype or teletype® (RTTY).

Another type of keying system is audio frequency shift keying or AFSK. AFSK uses the audio frequency shift to denote the "on" or "off" situation in the signal that carries the full data code. The data codes can be the common Baudot, or more modern ASCII systems, based on "0" or "1" expressions representing binary states or conditions.

In data language, the zero or one is referred to as a bit. Bit is a contraction of the word binary, which means something that is composed of or has two parts. Binary uses components such as transistors and flip-flops to read the binary information which is composed of series of bits arranged as a set code to produce the 26 characters of the alphabet. Each character has a combination of zeros and ones.

A five-bit binary code is known as Baudot Code. In Baudot, the on side or bit is known as the "mark" and the zero or the "off" state is known as the "space." Most Baudot teleprinters use 60 words per minute, 66 words per minute, or 100 words per minute—or expressed in baud rate would be: 45 Baud Rate = 60 W.P.M., 50 Baud Rate = 66 W.P.M., and 75 Baud Rate = 100 W.P.M.

Another common standard code is the American Standard Code For Information Interchange (ASCII). The ASCII Code uses 7 bits of information to provide 128 possibilities, plus an 8 bit for parity checking. ASCII is in common use for most data transmission and data processing.

It is a very simple matter to run inexpensive teleprinter circuits and systems on satellite subcarriers, with little loss in transponder output power.

FSK DATA EQUIPMENT (SbC)

Wegener series 2000 FSK subcarrier equipment is used to transmit data from 9.6 to 56 KB/S in the baseband spectrum above the video in satellite systems. The data interfaces are either RS 232C (for 9.6 and 19.2 KB/S) or CCITT V.35 (for 56 KB/S). Data transmission is transparent with no checking of protocols or error correction or detection.

Where full duplex operation is used, the data out of a demodulator can be switched to its mating modulator remotely. Data channels can be placed anywhere from 4.5 MHz to 10.0 MHz. Up to eight modulators can be summed in with the video by a high impedance loop-through in a single mainframe. Greater than eight modulators require the use of an external 8 way splitter/amplifier. This allows up to 72 channels to be placed on one baseband. On the receive end, up to 72 demodulators can be driven from a single 75 ohm source. **Figure 5.22.**

The Wegener Model 1606 Subcarrier receiver is designed to demodulate FM/FM/FDM multiplexed program, voice and data channels transmitted as subcarriers over video. The receiver is packaged in a standard Wegener series 1600 mainframe and is capable of feeding all series 1600 channel demodulators and accessories. The receiver consists of a down converter to receive the incoming 4 GHz signal, and convert it to a 70 MHz IF, a wideband IF amplifier/filter/demodulator to demodulate the video and subcarrier signals to composite baseband, and appropriate channel unit plug-ins. **Figure 5.23.**

The Model 1606 receiver features, single conversion down converter, wideband demodulator with thumbwheel transponder selection, precision AFC circuits for accurate IF passband centering, and the receiver interfaces with all series 1600 audio and series 2000 data demodulation equipment.

Turner Broadcasting System transmits a data channel with both CNN and CNN headline news (CNN/2) satellite feeds. The Wegener Series 2000 system provides up to eight independent 300 Baud ASCII information channels on a single low level data subcarrier. The data receiver/printer is designed to accept composite baseband video as an input and provide hard copy printout of messages associated with the Turner Broadcasting System feed. The data receiver is available in either the Model 2003–12 or Model 2004–12 configuration. The Model 2003–12

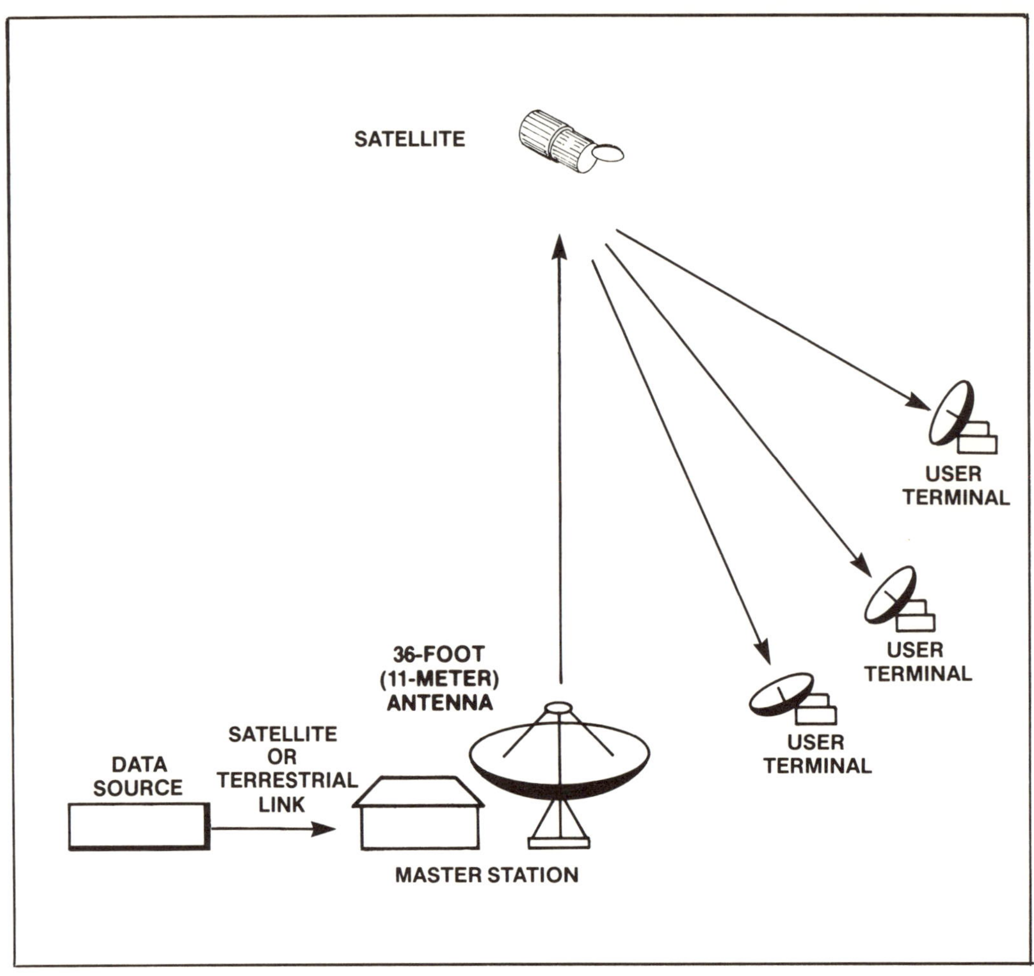

Figure 5.20 *Point-to-Multipoint data communication network using spread spectrum satellite link.*
Courtesy Equatorial Communications, Co.

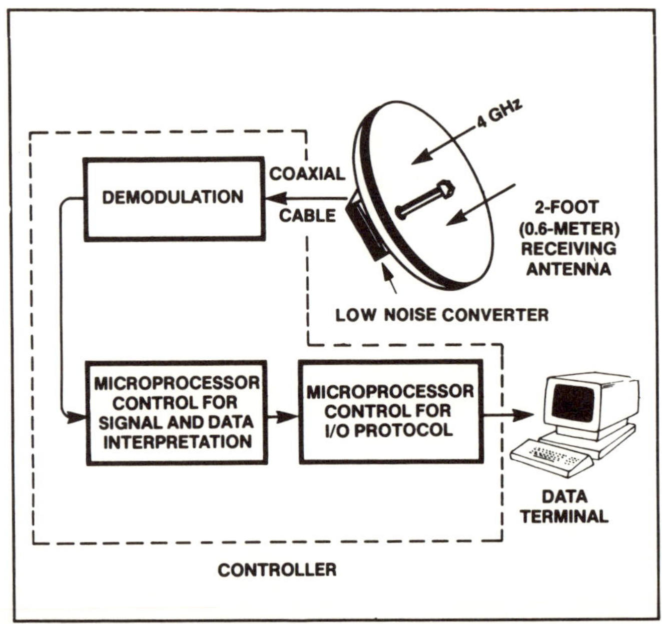

Figure 5.21 *Downlink. The satellite signal is transmitted at 4 GHz to the subscriber's two-foot antenna, amplified, and converted to an intermediate frequency for transmission by coaxial cable to the controller. The signal is then demodulated and sent as a standard digital signal under microprocessor control to user terminals.*
Courtesy Equatorial Communications, Co.

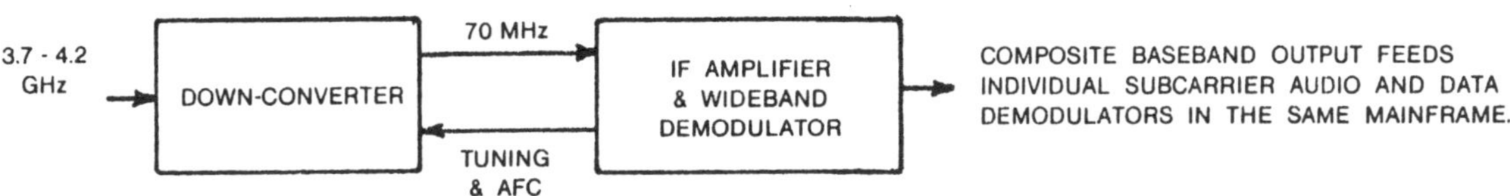

Figure 5.22 *Typical model 1606 application.*

Figure 5.23 *Wegener model 1606 subcarrier demodulator–receiver for FM/FDM–FM/DATA–FSK.*
Courtesy Wegener Communications

allows expansion capability while the Model 2004–12 has limited flexibility. The printer is an MX–80 by Epson. **Figure 5.24.**

Equipments features:

Precision low level subcarrier demodulator optimized to the Wegener Data uplink in use by Turner Broadcasting System, 3000 baud data system, at least 4 independent 300 baud printer channels are available on all models, expansion cards available for up to four more printer channels and unit will use Epson MX–80 printer.

Model 2012–01 Data Demodulator/Demultiplexer —The Model 2012–01 is a plug-in card that accepts composite unfiltered baseband video as an input and provides subcarrier demodulation of the 3000 baud data stream. In addition, the 2012–01 will demultiplex four independent 300 baud (ASCII) data streams providing RS–232C outputs for printers. The four 300 baud data streams are switch selectable on the card. (Any four of the eight available data streams can be selected.)

Model 2032–01 Data Demultiplex Expansion Card—The Model 2032–01 is a companion plug-in card to the Model 2012–01 above. This card accepts the demodulated 3000 baud data stream and demultiplexes up to four additional 300 baud multiplexed data channels. RS 232C outputs are provided, also switch selection of the data streams is provided on the card. These data channels are carried on both CNN transponders.

"SUN" OUTAGE—THE SATELLITE DATA NETWORKS PROBLEM

All good things seem to have a drawback, and satellite networks are no exception. The drawback is the phenomenon of "sun outages," which occur twice a year. These outages cause a severe degradation of the satellite signal for a period of time, and are due to the satellite's position in relation to the sun. These two occurrences can be predicted very accurately as to the time of day, duration and the degree of signal degradation that can be expected.

The primary concern is the period of maximum signal degradation which lasts only a few minutes. Other times during this period, signal degradation may not present a real problem to an efficient system. Most satellite networks have provisions for this brief period twice a year: Some go to telephone lines for proper program feeds; others simply shut the system down during the brief outage.

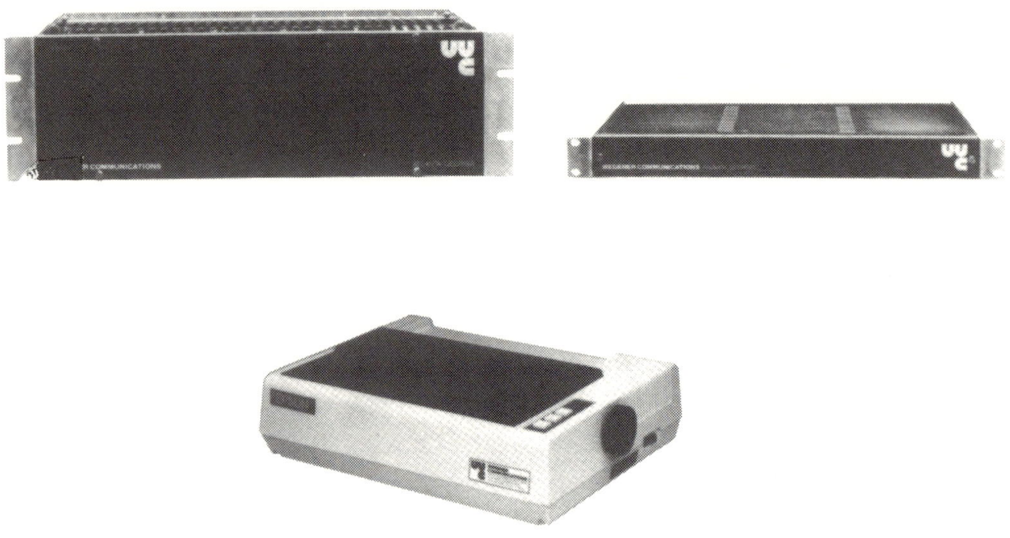

Figure 5.24 *Turner Broadcasting Companys' data system used on both CNN transponders. Data transmitted on low level subcarriers of CNN and CNN/2.*
Courtesy Wegener Communications, Co.

CHAPTER SIX

TELETEXT BASICS (TxT/VBI)

VIDEOTEXT: A PRIMER

VIDEOTEX—Generic term for systems that transmit text and graphics for display on a television screen.
VIDEOTEXT (VIEWDATA)—Interactive, two-way videotex service delivered via cable or telephone lines, allowing information retrieval and transactions.
TELETEXT—One-way videotex system that transmits text and graphics as part of the television signal over the air or via cable.
CEEFAX—British teletext system operated by the British Broadcasting Company. **Figure 6.0.**
ORACLE—British commercial teletext system.
PRESTEL—British viewdata system operated by government postal and telephone service.
TELIDON—Canadian teletext system.
ANTIOPE—French teletext system.
WORLD SYSTEM TELETEXT (WST)—A teletext format, based on the British standard, that is used in the United States and in more than 90 percent of all teletext systems in the world.
NORTH AMERICAN BASIC TELETEXT SPECIFICATION (NABTS)—A teletext format derived primarily from Canadian Telidon, currently in use in the United States.
CAPTIONS FOR THE DEAF (CFD)—The most-widely used teletext format in the United States, offering text to accompany television programming for the hearing-impaired.
VERTICAL BLANKING INTERVAL (VBI)—A generally unused portion of a television signal over which teletext and captioning can be broadcast.
FULL-FIELD—Use of a full television channel to broadcast text and graphic information. **Figure 6.1.**

ARE YOU READY FOR TELETEXT?

A number of television networks and satellite-delivered services are providing teletext services 24 hours a day, 7 days a week, but very few of us have decoded the contents. This chapter will tell you what they are, and most important, how to receive them.

The FCC defines teletext to be ". . . a new form of communication that involves the transmission of textual and graphic data on the Vertical Blanking Interval (VBI)—of the video portion of the TV signal.

If you are unfamiliar with these encoded transmissions, have a quick look. Tune in WTBS F3 TR6. Adjust the vertical hold control on the TV set until the picture starts to roll. See the white dots moving between the frames? *That is teletext.*

At this point, some additional terminology may be important. You may be familiar with the term "videotex." Videotex means all graphic and textual materials which can be displayed on a TV screen or monitor. Your kid's VIC 20 or VIC 64 probably can function as a videotex terminal, when connected to a phone line and with the proper passwords. The VIC can access data services such as the *Source* or *Compuserve,* or a couple of dozen other services, so your home PC (personal computer) can act as a videotex terminal. There are a couple of ways of getting that videotex onto your screen. One way is to hook your VIC or IBM, or whatever, to a telephone modem. The "modem" modulates or demodulates computer data which is transmitted by the phone line. Videotex usually connotes a 2-way system; the user's computer terminal transmits commands to the "host" computer. The host then retrieves the information and transmits it back down the phone line. **Figure 6.2.**

Teletext is simply a type of videotex. Think of it this way . . . the "tele" in teletext is descriptive of the mode of transmission, by tele(vision) or tele(communications). One other thing to remember: teletext, like television itself, or like a TVRO is basically a one-way affair. Information is received, but the user cannot transmit.

When most people talk about teletext, they refer to signals transmitted on the VBI of the TV signal, but this is not quite accurate. Text and data can be transmitted by any aural or visual medium. Hence you can transmit "teletext" by means of subcarriers, or digital audio encoding.

It might help to understand how teletext works; It is important to remember that all teletext is computer data. You probably already knew that computers operate on the basis of a coding system of zeros and ones. When the O's and 1's are deciphered by the electronic circuits, the computer can form numbers, letters, words, and most important colors and pictures. If you have dabbled in the world of personal computers, you will also know that while Apples, IBMs and Ataris can all do word processing (with the proper software), you can't shove an IBM diskette into an Apple and get the Apple to function with that IBM software disk.

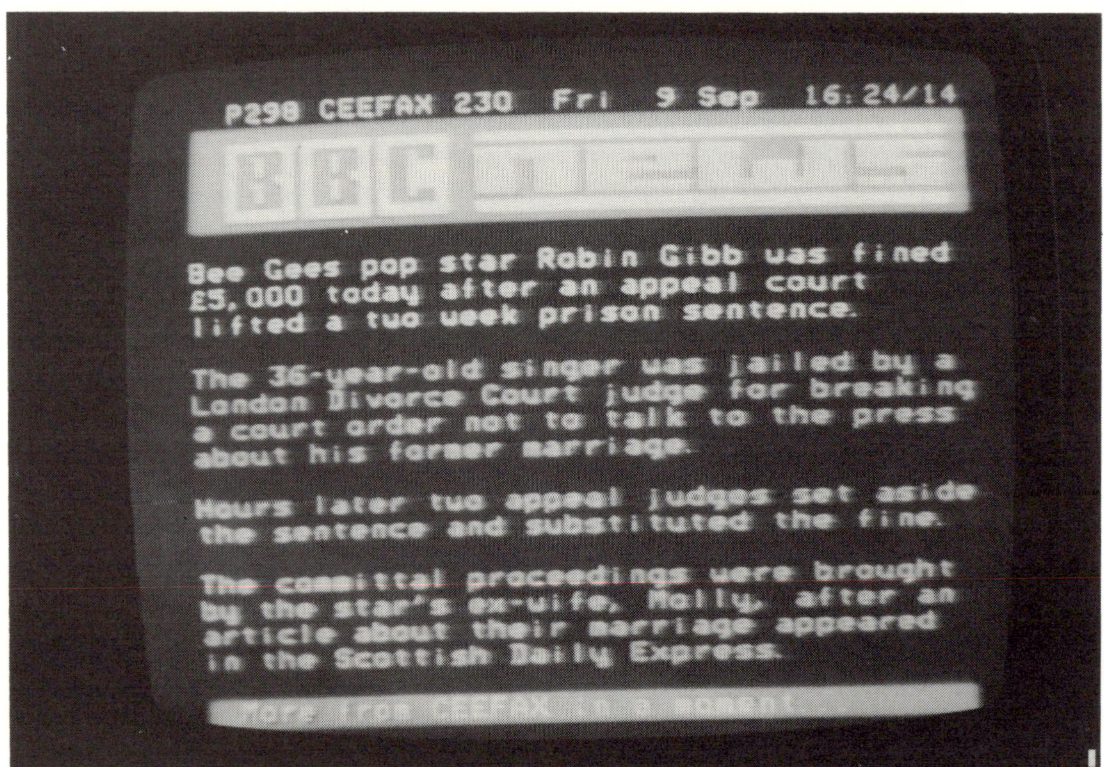

Figure 6.0 *CEEFAX, British Teletext System operated by BBC.*

Figure 6.1 *Zenith Teletext Decoder and Color Television Receiver will receive and display Teletext information.*

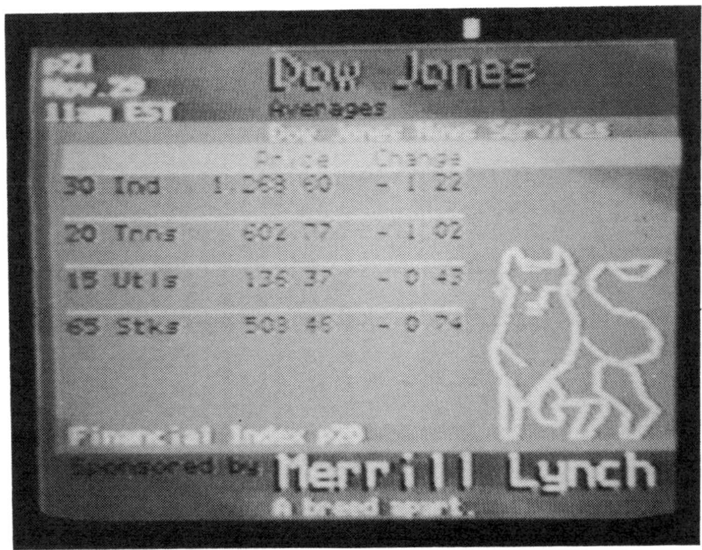

Figure 6.2 *Many services are transmitted via Teletex service (Dow Jones).*

Without engaging in too much technological discussion, let's just say that while personal computers are all based upon the same premise, they use different microprocessors, or different operating systems to achieve basically the same results. So it is with teletext. There are a number of systems now in the marketplace, and like personal computers, they have significant differences.

But before we jump ahead of ourselves with a history lesson, there are a few other things you must know. *All teletext decoders are personal computers.* They might not look like an Apple or Atari, but they contain integrated circuitry and in fact, they usually contain *more memory and circuitry* than the average personal computer! The software is usually inscribed right on the IC chips (RAM and ROM), so when you turn the teletext decoder on, it knows what to do; because the "software" is built-in (you don't have to insert a cartridge or floppy disk). Think of a teletext decoder as a personal computer designed for one purpose and one purpose only, to receive information "pages" via transmitted signals. When we talk about the pricing of teletext decoders later in this article, you will recall that you are getting a sophisticated personal computer.

A SHORT HISTORY OF TELETEXT

More than a dozen years ago, television engineers started to realize that they could cram all sorts of *digital* information into the vertical blanking interval (usually lines 13–21) without physically interfering with the TV picture. At first, the BVI was used for special monitoring signals which allowed engineers to perform quality-control checks on the TV signal as it passed from the network studio to microwave links and ultimately to broadcast transmitters. The public got involved with the VBI signals in the late 1970's, when several TV set manufacturers took advantage of VIR (Vertical Interval Reference) signals which were transmitted on the VBI's of various stations. The VIR signals were used for quality assurance and color fidelity. Several TV receiver makers, including General Electric, reasoned that a set locked into the VIR signals would produce better color pictures, more consistently, on more channels. VIR signals, it was thought were not as prone to multipath interference as the traditional color control signals. **Figure 6.3.**

Also in the 1970's the Public Broadcasting Service in conjunction with the National Captioning Institute (NCI) and Texas Instruments, developed a system for close-captioning of TV programs for the hearing-impaired. This was a very crude system of teletext. Signals were transmitted on line 21 of the VBI. The viewer purchased a decoder (sold by Sears) or a receiver equipped with a built-in decoder. The decoded captions were superimposed over the picture in the lower portion of the screen.

Closed captioning was a crude form of teletext because the graphics capability was quite rudimentary. The captioning was usually displayed as a one-line message scrolled along the bottom of the screen, similar to a computer print-out. The captioning materials usually paraphrased the dialogue of the television program. Word-for-word translations were impractical because the average reader could not comprehend the one-line caption at a rapid rate.

Three networks opted to broadcast closed-captioned programs *PBS, ABC* and *NBC.* CBS did not initially opt to use the NCI system because CBS engineers believed that this system was inefficient, and it used up a significant portion of the Vertical Blanking Interval for *low-resolution* graphics. CBS reasoned that if NCI closed captioning proliferated, it would be difficult to introduce a *high-speed* data-transmission using high-capacity transmission standards.

By 1978, there were a number of teletext systems under development throughout the world. The British had developed a system (called *CEEFAX,* or *ORACLE* or *PRESTEL),* which was based upon alpha-*mosaics.* The Canadians had invented a system (called Telidon), based upon alpha-*geometrics.* What's the difference? The British system created shapes in much the same manner that you might create a drawing on a computer. On your computer terminal, you can create graphics by typing in the right combination of characters and dots. If you have a computer terminal, you have probably seen the program which will draw a woman by means of X's and O's. That's a very crude form of alpha mosaics. The Canadian system was more dependent on the computer to assist in drawing pictures. The difference in the two systems is rather striking; The Canadian system could create circles, arcs, and all other geometric shapes. This meant that artists could convert realistic drawings into teletext pictures. The best the British could do was create pictures based upon squares, boxes and jagged lines.

The French developed a system called Antiope/ *DIDON* which share attributes of both the British and Canadian systems. The British, French and Canadian governments poured a lot of money into their respective systems, so an international standards battle broke out. *Each country* wanted *its system* to be adopted as a *"world standard."* Each country knew that the real victory would occur in the United States, where there exists a market of more than 86 million television homes in an information-hungry nation. No one wanted to repeat the standards battles of the 1940's and the 1950's, which resulted in a lot of different television transmission standards.

The French and Canadians found some common ground but the British were unable to agree on anything, *except that* they wanted everyone to adopt *their standard.* In 1980, several manufacturers with support of their respective governments, supplied a number of American broadcasters with prototype equipment to conduct field trials. *WETA* Washington, D.C. used *Canadian-made* equipment, *KMOX* used some *French* equipment, and *WKRC–TV* Cincinnati, Ohio (no relation to the radio station in the sitcom TV show) used *British* equipment. NBC used French equipment at KNBC. There was a lot of lobbying in Washington, as each faction tried to get the FCC to move toward *one teletext standard.* (Reminiscent of the Quadraphonic and AM Stereo battle, isn't it?)

You have probably noticed that we haven't mentioned an "American system" of teletext or a "Japanese system." There was a Japanese system developed called *CAPTAIN,* but it seemed to fizzle out in the face of other international competition. Most American and Japanese companies sat back and watched the other governments and their engineers fight. They figured that eventually they would be involved in the final outcome, and they were right. Committees were formed and a number of American companies including the giant AT&T, and CBS and NBC, set out to determine a "protocol" for transmission of Videotex and Teletext. Over a period of two years, a transmission system was developed which incorporated a lot of the original Canadian Telidon system, along with elements of the French Antiope & DIDON systems, and a lot of new twists thrown in for good measure. This meant that much of the equipment used on those original field trials had to be scrapped as the standards evolved.

The new protocol, known as *NABTS* (North American Broadcast Teletext Specifications) sets down a basic blueprint for teletext transmission, although individual manufacturers or broadcasters have made some minor changes of their own. Meanwhile, during the time that AT&T, CBS, NBC, Time, Inc. (the parent of HBO, and Cinemax), and others were sitting around the bargaining table, the British started to make inroads in the U.S. with individual broadcasters.

NO DATA HERE/standard frame bar has 'softer,' less 'busy' appearance.

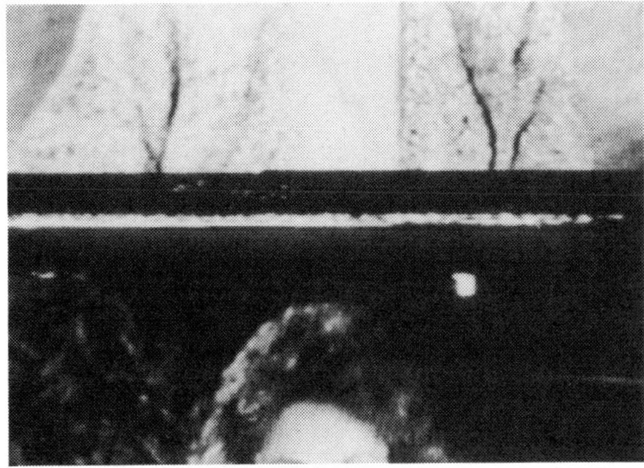

TELETEXT/Roll up (or down) your vertical hold and 'freeze' the sync bar near the middle of the display screen. Along the bottom edge of the bar you will see 'dancing' dots of data; containing the Teletext message material (CBS on D3, WTBS on F3R, etc).

Figure 6.3 *VBI on TV Set*

Many teletext observers wanted the FCC to declare that there would be only one technical standard for TV transmission, but the FCC decided to authorize teletext under the "open market" approach. Like AM–Stereo, the FCC decided that the only restriction should be *non-interference with* the regular broadcast service of the originating station.

THE SATELLITE-DELIVERED SERVICES:

At the time this book was written, the major services on the satellites are: Keyfax, located on the WTBS VBI F3, TR6. Time Teletext *has been found* on F4, TR18 and F2R TR23 up until December 31, 1983. Time, Inc. has wound up its experimental service, however don't be surprised if you see the Time Teletext signal up and running on various satellites from time to time. Many people thought that the signal on F4 TR18 was a form of scrambling for HBO, because digital data which did not really resemble the Reuters transmission system occupied the entire transponder on F4. "Extravision" is the CBS service which can be found primarily on Comsat D3. NBC's teletext service can be found on Comsat D3 TR1. It is reputed to be based upon the NABTS system of teletext, but we have never been able to decode it, despite the fact that our decoder works on all other teletext services. **Figure 6.4.**

More teletext can be found on Anik B and Anik D, on the CBC channels. The English language channels have a service originated out of Toronto, while the French language channels have a teletext service originated out of Montreal, although on Anik D, we have sometimes found French language test slides running on the CBC English language transponder, which originates out of Montreal. TV Ontario on Anik C3 can be found from 8:30 am EST to 11:30 pm EST with a news and information service. The hours are sporadic on weekends. **Figure 6.5.**

MAJOR TELETEXT SERVICES

What is the content of the major services? Let's have a look. *KEYFAX is the first commercially available service.* It is transmitted on F3 TR6 on the Vertical Blanking Interval of WTBS. The KEYFAX system uses British "CEEFAX" technology which is sometimes referred to in the advertising literature as World Standard Teletext (W.S.T.). However, as we mentioned, *there is no* "world standard" (the British is *one type of* teletext, not a standard). It is an alpha-mosaic system. **Figure 6.1.**

The content of the service includes world and national news, sports scores and business news. There is also weather news, and a newsflash feature, which allows you to have bulletins displayed on your TV set even while you watch WTBS. Keyfax also includes leisure and entertainment features, and will *probably contain* commodity information later.

The KEYFAX decoder (see **Figure 6.6**) is made by Ayr Electronics of Great Britain. There is an infrared keypad to call up individual pages. When the decoder is turned on, a menu page is displayed. On the menu page there are numbers which refer to features; NEWS might be found on pages 001 to 019. To receive the news, you would "key in" 001.

When I contacted KEYFAX, their marketing strategy had just been modified. Originally, they were only looking at marketing through cable companies; however, *now it is possible* to purchase a decoder. The pricing of the decoders is around $300. The suggested retail price for the service via cable is $19.90 a month, with half of the fee for content, and half for the leasing of the decoder. Eventually there will be sponsorship of pages in the service. The marketing executive for KEYFAX Teletext is Selman Kremer.

Mr. Kremer works out of the SSS office in Tulsa, Oklahoma. SSS is a partner in the service, with Keycom of Chicago.

By the way, *in England* the BBC is transmitting *computer software* to home personal computers using the same VBI signal that contains its CEEFAX. It is not known whether KEYFAX intends to transmit computer software, but the technology will permit the transmission.

EXTRAVISION

CBS has made a significant commitment to teletext. Their service is known as EXTRAVISION and consists of about 50 pages. Unlike KEYFAX, CBS uses the North American Broadcast Teletext Specification (NABTS). That means that the Keyfax decoder will *not work* on CBS. The CBS service has generally the same content as KEYFAX, with a couple of differences. CBS has some fascinating pages related to the communications and entertainment industry. It also has a few pages for the hearing impaired and a listing of programs for the "4 major" networks. On the menu pages, there are always 2 news headlines which

Figure 6.4 *Many people thought this TIME, Inc. service on F4/TR18 was scrambled television.*

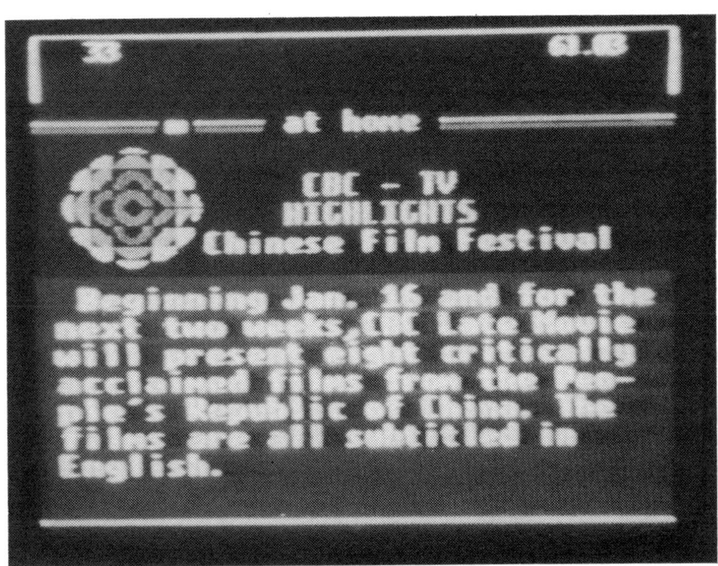

Figure 6.5 *CBC Service makes excellent use of background colors (ANIK B and D) and creates easy-to-read format even on a small screen display.*

"blink." Almost every page on the CBS service contains advertising, and despite the lack of decoders in TV-land, there does not seem to be a shortage of advertisers on Extravision. Advertisers include Coke, Ford, and Phillips 66. I happen to like the CBS service, primarily because the pages flash up on the screen quickly (more about the speed of the service later). The other nice thing about CBS is that the service is free of charge. **Figure 6.7.**

The NBC service has been up and running on D3, TR1. It is reputedly in the NABTS format, but there seems to be some problems with the method of encoding. Our *Norpak* decoder would not lock onto NBC. One engineer theorized that NBC was using some French software, and the NBC database was structured differently than most NABTS databases, which meant that when the decoder was turned on and automatically searched for the "menu pages," it in fact searched for page numbers on NBC, which simply were not in their databank. Consequently, the decoder would not "lock on" to the NBC service. We are trying to get more information from NBC concerning the problem.

Time Inc. (parent company of HBO and Cinemax) conducted the most expensive and extensive test of teletext. Their service occupied a full transponder (not just the Vertical Blanking Interval) and they had the capability to transmit up to *5,000 pages* simultaneously(!). When the service was good, it was very good. It was primarily delivered via F4 TR18 and F2 TR23 to cable head-ends in Orlando, and a couple of other cities. There, test homes were equipped with decoders, which had the added feature of "downline loading." That meant that the software for video games could be loaded into the decoders.

The Time service also had TV listings, horoscopes, travel information, sophisticated weather data, even an astronomy section. The potential was there, but Time, stunned from its multi-million dollar losses in a weekly cable-TV listings magazine, pulled the plug.

The Canadian Broadcasting Corporation has multiple teletext feeds on Anik B & D. The French language channels have French teletext originated in Montreal, while the English language channels originate in Toronto in English. The news, sports and weather service is pretty good. The CBC is way ahead of CBS in their use of graphics. For example, with the CBS service, you'll strain your eyes if your TV set isn't a *monitor.* The CBC uses a combination of background colors and type fonts to overcome the problem of print on the electronic screen. Recently some advertising has crept into the CBC service. One of my favorite pages was an ad for Baskin Robbins Ice Cream. One of the other CBC features is an automatic "page advance." You don't have to hunt for pages, the machine changes pages at the rate which an average reader would. Shortly, the CBC will add captioning and headline flashes to its teletext service. As of January 12, 1984, closed captions were running on some teletext channels, but the captions had nothing whatsoever to do with the program content, they just seemed to be testing the captioning equipment!

You may have noted some cryptic references to the "NABU NETWORK" in columns devoted to "future satellite services." The concept is similar to teletext. A microcomputer having 80K memory receives its software by means of a digital signal modulated on an RF channel. The data is processed through an addressable modem. That means the computer owner can purchase "tiers" of software. At the present time, the NABU computer is available commercially in Ottawa and Vancouver (Canada), and will shortly be launched in Alexandria, Virginia. In those locations, the computer will be fed by means of the cable television system. However, last spring *NABU* demonstrated its system using the Anik B satellite. At present more than 35 computer programs including video games (PAC MAN, POLE POSITION), word processing, stockmarket quotations and wine guides are being transmitted, and shortly the inventory should exceed 100 computer programs. When you consider that the monthly charge for a single tier of computer programs is roughly the same as the retail price of HBO, and when you compare the cost of purchasing computer programs, you will realize that *the NABU people are on to something which could revolutionize the computer industry.* There are rumors that their system may soon work with the IBM PC. By the way, the computer system sells for *under $600* and is very similar to the new Japanese MSX computer standard. Computer software by satellite? Perhaps sooner than you think.

Now that you have a fairly good idea of the types of teletext services now available and their content, you may wish to receive teletext via your TVRO. The list of do's and don't's is not exhaustive, but there are a number of important considerations.

AVAILABILITY OF DECODERS

First, the availability of decoders. As mentioned previously, the service telecast on WTBS is available only with a WST-type decoder (model T100) which is manufactured in Great Britain. It is available for $399, from SSS in Tulsa. The man in charge is Selman M. Kremer, (he is Executive Vice President of Satellite Syndicated Systems, Inc., Box 470864, Tulsa, Oklahoma, 74147; telephone 918–481–0881.) *One caution:* they are *not* particularly keen to conquer the TVRO market. They would *rather* market through CATV systems. Their target is 1% of WTBS-cabled homes within 3 years. While that is obviously a large market, the TVRO market is also significant. With some persuasion, they might become more interested in the TVRO market. But, "rumor" has it that the decoders which are now being sold in Cincinnati and *made by Zenith* for Taft Broadcasting's "ELECTRA" teletext service (transmitted on WKRC-TV) will work very well on WTBS. Best of all, they are being sold in Cincinnati for $300. There are 2 models; the DX–1000 and the DX–900. Apparently the model 900 was designed specifically for KEYFAX. There has been some suggestion that the Zenith decoders must be used in conjunction with late model Zenith color TV's, but I am certain that some enterprising TVRO owner might be able to get the Zenith decoders to work with other TV models, with a simple plug change for proper hookup.

The decoders for North American Broadcast Teletext Standard (NABTS) which is transmitted by CBS, NBC, Time Inc., CBC, TV Ontario and Learn Alaska are at the present time *expensive,* and in relatively short supply. You'll recall in the introduction to this chapter, we mentioned that teletext is a computer system. The NABTS system of teletext uses high resolution graphics. Therefore, in that decoder box is the front end of a TV receiver, plus filtering devices to differentiate the VBI lines, *plus a digital tuner* which will tune VHF and UHF and mid-band and super-band Cable channels, and a computer to decode the data, plus an infrared remote system, plus RGB circuitry for display on a high-resolution monitor, plus a modulator. *All of that adds up to a lot of money.*

I am using a decoder made by *Norpak Corporation* of Kanata, Ontario (an Ottawa suburb). The decoder retails for about $3000. As you can imagine, that price is enough to cool anyone off quite quickly. But the truth is, quantities of teletext decoders have not been large to date. Also, the two year negotiated battle over transmission and display standards meant that R&D was continuous, and terminals which should have been production models became prototypes. Because there was something of a metamorphosis of the technology, a lot of extra circuitry had to be built into the boxes. Likewise, these were never really meant to be consumer models; most have been designated for "field trials." In summary, the situation concerning NABTS teletext decoders is somewhat analogous to TVROs in 1979; limited production runs, sometimes unreliable, a variety of different circuits, high consumer prices. *But don't count NABTS teletext out yet.*

Several things are about to happen: First, a number of major manufacturers are working on VLSI. That stands for *Very Large Scale Integration* of circuitry. Most major functions will be integrated on a chip set. That means there will be a major price breakthrough. Since one chip or a couple of chips will perform the function of many circuits, reliability should be improved. Rockwell International has joined with Norpak of Canada to design VLSI. The Japanese companies led by Mitsui are also interested; as are AT&T, RCA/NBC, and a number of other major players. It is rumored that Texas Instruments has a chip set nearly ready which can be built right into a microcomputer. Manufacturers are also getting ready to integrate teletext into TV sets. If the trend which I noted in England last summer carries over to North America, Sony and the other component TV makers will offer "component" teletext. In other words, the market for teletext is about to explode. By early 1985, cheaper decoders using VLSI could be moving into homes.

There are a number of technical problems which will impact on TVRO systems, and if the TVRO industry doesn't get involved, and quickly, there will be problems, mainly from disgruntled consumers who have purchased TVRO systems in good faith for teletext use.

First, let's outline the equipment which we used for our tests: The receiver is a Dexel DXP1102. It has an LNC, which was tested at 93K. The dish is a Hastings 10½ footer with metal panels. Reception was rated at "sparklie-free" to the odd sparklie on the test channels. We were taking the modulated video out of the Dexel on Channel 3. Although the Dexel has a "video out" port, which they claim is filtered (it isn't), the teletext decoder didn't have a video input port. So

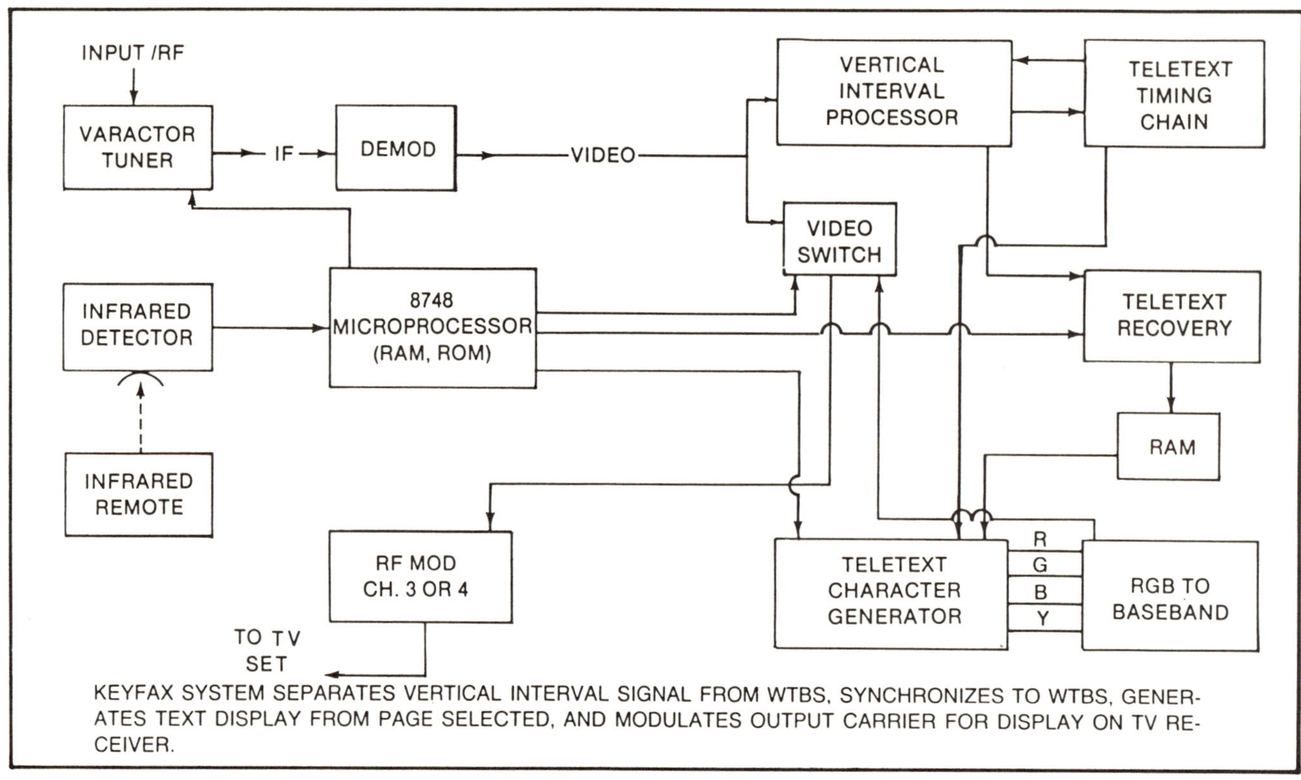

KEYFAX SYSTEM SEPARATES VERTICAL INTERVAL SIGNAL FROM WTBS, SYNCHRONIZES TO WTBS, GENERATES TEXT DISPLAY FROM PAGE SELECTED, AND MODULATES OUTPUT CARRIER FOR DISPLAY ON TV RECEIVER.

Figure 6.6 *Keyfax System separates vertical signal from WTBS, synchronizes to WTBS, generates text display from page selected, and modulates output carrier for display on TV receiver.*

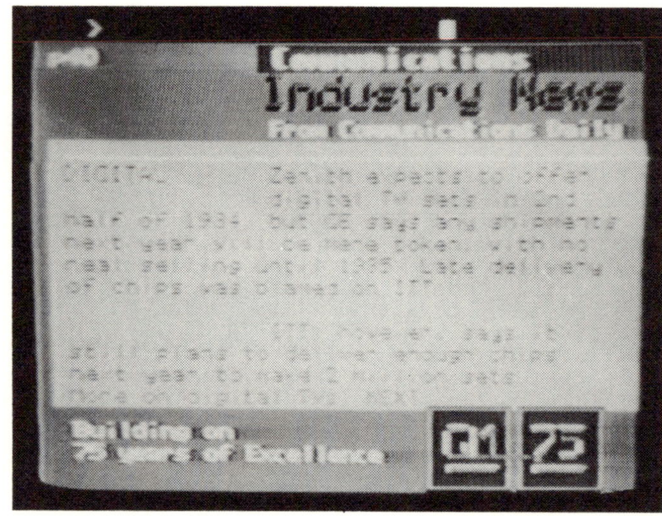

Figure 6.7 *CBS Extravision (TR10, D3) is excellent service covering basic essential data.*

there was *the potential* for some clipping of the VBI *in the Dexel modulator.* We would have preferred to run video directly into the decoder. We were able to run the demodulated video output from the Dexel into a Panasonic NV–8250 industrial model VHS machine, and through its modulator tuned to Channel 3, into the teletext decoder. Our tests showed that there was little difference in performance of the teletext decoder between the 2 modulators.

HERE ARE THE PROBLEMS

The VBI information, which appears as a bunch of "flashing dots" in the vertical frame-bar, are a *precise coding system.* The teletext decoder recovers computer data from those "flashing dots." Then the data is processed; the data instructs the teletext unit where to draw lines, how to color the pictures, how to display text.

Earlier in this chapter, we made reference to waiting times. Try to visualize teletext as a large magazine which is transmitted page by page, over and over again. If you turn your decoder on, and want to see page 62, you have to wait until that point in the transmission cycle when page 62 is transmitted before it can be displayed. In theory, a teletext magazine on the VBI could contain hundreds, or thousands of pages of information. *However,* if there were hundreds of pages in the service *you would have to wait a long time* before the page you requested would be transmitted in the cycle. In other words, because VBI teletext only occupies a small part of the video bandwidth, you compromise the size of the data base in favor of cycle time. Most VBI teletext services contain less than 200 pages. To keep the waiting time for viewers down to a minimum, key pages (such as the menu page) are included more often in the transmission cycle, so there is a faster response time. This is why the CBS service contains approximately 50 pages and CBC contains about 100 pages. Time Inc., used a *full video channel,* with 525 lines devoted to teletext. As you might have guessed, they had hundreds of times more capacity to transmit pages to their viewers, and the transmission cycle was much faster than VBI teletext. **Figure 6.8.**

We also mentioned that there were problems with satellite transmission of teletext. To better comprehend the problem, remember that teletext is a computer transmission system. Instead of using phone lines or wires, teletext is transmitted on TV pictures. Most computer systems perform quality control functions to ensure that the data received from a remote location is accurate. Teletext is no exception. *"Error rates"* have been established. This is an important aspect of the NABTS teletext transmission system (and all other data systems for that matter). That means the decoder performs certain verification tasks. If there are too many anomalies, the decoder or computer won't display the data. In other words, the machine virtually shuts down if there are significant errors. You may recall earlier concerns about the Reuter's system. They had stringent standards concerning signal quality, and had reservations concerning allowing TVRO owners to access their system for fear of greater than acceptable error rates. We have the same problem with teletext.

1) *Sparklies.* This is simply the number 1 problem facing teletext. The sparklies are displayed in the video signal almost identical to the digital data in the VBI. While the decoder can differentiate the VBI data in lines 13–21 of the picture, from the other video information, the sparklies appear in all lines of the picture. *That means if you have sparklies in the picture, you have erroneous information in the VBI.* While you and I know that the sparklies aren't supposed to be there, and we might ignore the sparklies, the decoder thinks that they are data and instructions for displaying pictures. With sparklies, you get errors in the data, and that can cause the decoder to reject all of the data, and hence, no teletext picture.

This phenomenon concerning sparklies is not limited to teletext. Recently, I had a look at the Scientific Atlanta receiver for the new NBC (and other networks) *digital radio* transmission system. There was a very strong wind buffetting the dish which was on the roof of a radio station. The LNA was moving. When sparklies (noise) were introduced into the signal, the audio did not get noisy, *it simply got cut off!* With digital transmission, the general rule is, if the data gets through, the reproduction should be perfect; if the error rates are exceeded, *you get nothing!*

Unfortunately, one of the trends in the TVRO installer business seems to be that of misleading people. Recently a friend of mine was shopping for a TVRO system for his home. A majority of the sales people (some who had been in the TV business for a lot of years and sell on the basis of their fine reputations) tried to sell undersized antennas with LNAs that weren't fit for the job. They tried to con him that sparklies on F3 were a normal fact of life; something that came with TVROs, something you lived with. *Unfortunately,* a few sparklies will virtually kill chances of decoding teletext, now and forever.

If your clients are potential users of teletext (and I would think that most TVRO owners will want TxT, then the dealer/installer has a moral responsibility to make sure that customers purchase and receive sparklie free signals.

When we tried to decode signals with even moderate sparklies, we either got pages with garbled data, or nothing at all . . . the machine would not display pages properly when sparklies were present.

2) *Saturation.* I would rank this problem second to the sparklies problem. Unfortunately the Dexel DXP1102 is notoriously bad on saturated colors. Despite blistering signals from D3 and Anik D and Anik B, we had a lot of trouble. The saturated reds on the Dexel display as a form of sparklies. Although the saturated colors are in lines of the video signal *other than* 13–21 of the video signal (where teletext is transmitted), for some reason, the decoder was "fooled" and we either got garbled data or no data at all. *Game shows on CBS were the greatest problem.* The Price is Right (for example) uses a lot of saturated colors, a lot of chroma key, and a lot of blinking lights around displays. I literally had to wait until commercial breaks to get the teletext decoder to display CBS pages. One small consolation; during prime-time there aren't as many shows with saturated colors, so you don't run into this problem as often. *But on Magnum P.I. they seem to wear a lot of Hawaiian shirts with saturated reds!*

I understand that a lot of companies are abandoning the 564 circuit, and now are addressing the saturation problem. The industry rule should be, *if there are no sparklies, there should be no saturation.*

3) *Modulators.* This did not seem to be a severe problem with the Dexel, but it is a potential problem for teletext. The receiver modulator must not only pass the lines in the picture portion of the video, but it *must also pass lines 13–21* without distortion. Unfortunately some manufacturers skimp in this area. If you are a manufacturer and use a brand-name VCR modulator, *you should not assume* that the modulator was designed to pass the VBI. *Some don't.* You can also be easily fooled. While you might think you see VBI signals coming out of the modulator, enough distortion may have been introduced at the time of modulation to create significant errors. As noted, the Panasonic modulator passed through the VBI data but the videotape machine could not properly record and reproduce those VBI signals. Ironically, while we found that the Panasonic VHS machine seemed to record signals containing VBI data, and on play-back, all of the pulses seemed to be in the VBI, we could not recover the data. The reason? *½" videotape machines do not record or play back all of the lines.* While the picture may appear to be 525 line NTSC, it is in reality a degraded signal with a loss of resolution. That resolution comes in the form of lines which are displayed.

The engineers at Norpak indicated that they even had some problems recording VBI data on *¾ inch* U-Matic machines which comply with broadcast standards.

4) *Videoprocessing.* As we mentioned, we were not able to evaluate raw video going right into the teletext decoder simply because the Norpak unit did not have an input port. But in the future, *TxT units may have video inputs.* Obviously, the less you process a signal containing VBI teletext, the better the odds that you will not introduce noise or other aberrants into the signal. The problem would seem to be a generous interpretation by manufacturers of "standards" for video output on their receivers. For example: Dexel's specifications claim that the video output is filtered and clamped. Although we did not run the output through a scope, we were able to recover all audio subcarriers from 5.3 to 8.1 MHz when the video output was run through an audio demodulator. That meant that a lot of what was reputedly filtered, got through. Also, video amplification is achieved through 2N2222 transistors. We had some problems with the processed video. Eventually one of the videoamps had to be replaced. We put in 2N2222A transistors, which are a little faster in switching. There was a significant improvement. Somewhere along the line, *receiver manufacturers had better awaken to the fact that consumers will eventually plug computers,* and teletext decoders, and monitors and yes, "legal" descramblers, *into their satellite receivers.* Manufacturers had better build in proper video ports and RS232 connectors. There should be a filtered port, and a port for raw unfiltered video. Likewise, more care and attention should be devoted to the actual levels which come pouring out of receivers. In my experience, too many TVRO systems compromise video performance and consequently, what comes out of a receiver bears little resemblance to NTSC or any other "standard." In short, *if data of any kind* is transmitted via satellite,

be it in subcarrier form, or encoded video, *the consumer should be able to recover that data without distortion.*

One other point: Sometimes consumers or manufacturers use RF amps to amplify, and then split the modulated signal coming out of the receiver and feed multiple TV sets. One must also be careful with the RF amps. In our experience, while you might get acceptable pictures out of the amp which seem to be noise-free, you might also distort the VBI with RF amp "noise."

5) *Audio quality.* Some teletext is transmitted by means of encoded tones (i.e., Genesis Story Time). Many receiver manufacturers have adopted audio sections in their receivers, which make most audio engineers cringe. Until recently, I think most receiver designers have placed audio processing at the bottom of the list of priorities. I have personally seen many receivers which have questionable audio frequency response. The *early* Sat-tec receivers suffered from a poor audio section, which was subject to severe distortion on high frequencies.

Very few receivers can accurately reproduce audio above 10 kHz. Many receivers have audio response which may contain some top end or bottom end frequencies, but are far from linear in response. The irony is, if an LNA manufacturer put out a product which deviated 10 dB here, 15 dB there, or had high levels of noise, the receiver makers and distributors would yell and scream. Unfortunately there is a double standard in the receiver industry because the audio sections of receivers often deviate many dB, and are just plain noisy; yet little has been done and there has been little public discussion of the problem. There are some receivers which distort badly on narrow band signals. Others, like the Dexel 1102, advertise subcarrier tuning, but cannot tune the subcarriers on WGN or CBN or WTBS. The Dexel audio section was designed for MTV and the Movie Channel stereo, and 6.2 and 6.8 main channel audio, but just won't work properly on other subcarriers. We wrote the manufacturer, but no one even bothered to acknowledge our letter. There are signs that this is changing as some manufacturers have realized that stereo and high fidelity are major consumer inducements. (Even Dexel seems to be addressing the audio issue in its new receivers.) But the truth is, *many receivers can't tune in subcarriers properly.* In our complaint about video processing, we noted that many receivers have filtered video outputs, so even if you wanted to add an audio processor, you would have to cut right into the circuits of the receiver. The real answer is to provide a wide-band output, and more care and attention must be paid to audio processing, with proper equalization and flat frequency response. One other point: many modulators also skimp on frequency response and that is another reason for providing audio outputs.

If you are a manufacturer of TVROs, you are probably wondering how can you obtain a NABTS teletext decoder to ensure that your receivers will perform properly. At present, Norpak Corporation of Kanata, Ontario makes a teletext video decoder which is a rack mount unit. *It sells for $4,360(!).* They also make a dual mode videotex and teletext decoder which is somewhat more versatile. It sells for approximately $3,000. Apparently, the dual mode decoder does not have FCC type approval, so you might have some difficulty getting one across the border, legally. One of the salesman at Norpak is Bob Croll, and he can be reached at (613) 592-4164.

To sum up. In the next 5 years, teletext will become a multi-million dollar industry. Electronic publishing has the potential to cater to every taste. Since the cost to transmit VBI signals and create pages is under $100,000 (the cost of 2 TV mini-cameras), you can almost be certain that there will be dozens of services up and running shortly. There will be some fascinating partnerships formed between traditional "broadcasters" and print publishers. CBS and NBC are poised to supplement their revenues with teletext. Obviously, people who demand the latest entertainment technology will be likely candidates for teletext. *People living in remote areas,* unable to obtain print material easily *will also demand access to teletext.* As digital audio, computer software transmitted by satellite, and data services are introduced into the home, consumers will demand sparklie-free and saturation-free signals. Now is the time for the industry to take action.

KEYFAX TELETEX SERVICE

The operation of the keyfax service is outlined in the block diagram in **Figure 6.6,** and can be described in the following manner: The TVRO is tuned to SATCOM IIIR Transponder 6 (WTBS). The keyfax data is carried on the vertical blanking interval. The RF is demodulated into baseband video, and passed through a chip (the Vertical Interval Processor) that separates the data contained in the Vertical Blanking Interval of WTBS. The baseband video is made available to a solid state video switch, that also accepts the teletext video that is about to be generated.

The processed video from WTBS is then placed through a timing chip that maintains synchronization between the teletext decoding circuits and the incoming WTBS video. The data is then extracted from the Vertical Blanking Interval in serial form and is compared to the page number selected. If the page numbers being transmitted match, the data is decoded and placed in parallel form into the teletext character generator. The character generator's output is in the form of R, G, B and Y signals. These signals are reconstructed into a baseband video signal that is applied to the other side of the video switch, previously mentioned. The output of the video switch, either WTBS or keyfax, is then RF modulated, and applied to the antenna input terminals of the subscriber's television.

The present keyfax decoder has the ability to grab whatever is on the "VBI"—limiting the inventory to about 100 pages. It is expected to increase page count by 1985–1986, by the use of a home computer. This would also allow transmission of the most sophisticated video games, information storage, and the ability to address specific subscriber information.

For the data now being offered on keyfax, there are contracts with Associated Press, United Press International and Dow Jones Services, plus information out of the Chicago Sun-Times computers. In Chicago, a staff of some 25 persons is involved in rewriting the information in the keyfax format and providing graphics for the system.

A few of the features of keyfax's offerings are: News—fast, accurate and factual reports on international, national and regional stories updated 24 hours a day.

Sports-in-progress reports on all major games plus results, statistics and standings from across the country are included with previews, personality profiles and commentary.

Business—From stock quotes to home finance tips, the business section follows all the action on Wall Street, the New York Stock Exchange, AMEX, COMEX, dollar and gold markets, plus analysis of fast-breaking financial stories.

Weather—Complete regional weather reports presented through colorful graphics and forecasts.

Special features—Regional features on the people and events that are top priority. **Figure 6.9.**

Keyfax service is now offering a new mini-magazine of news features and other information for hearing-impaired persons. This service is in addition to the "speak-up" service, and offers information contained primarily in the National Captioning Institute's Hearing Impaired News Text Service (HINTS), which includes stories and features on legislation affecting hearing-impaired persons. Up-dates on programming and news text summaries are also provided.

Keyfax decoder units are to be made available through several sources—write Southern Satellite Systems, Inc., P.O. Box 45684, Tulsa, Okla. 74147.

ZENITH TELETEXT SERVICE

Zenith began research and development in teletext in 1978, becoming one of the first American electronics manufacturers to actively pursue the new communications medium.

Here is a brief synopsis of Zenith's research and development efforts in teletext: **Figure 6.10.**

1978. Early broadcasts of teletext were originated at KSL-TV in Salt Lake City, Utah, and KMOX-TV in St. Louis, Mo., where Zenith color TV receivers were equipped with teletext decoders, based on the proven British standards.

1979. To investigate the many possibilities of videotex, Zenith engineers built a special computer-controlled display generator that has been widely used as a part of the industry development work at Zenith.

Zenith was approached by Satellite Syndicated Systems, Inc., (SSS) in its search for teletext decoders suitable for use with a satellite-delivered newswire service, now called CableText, in the United States. Company researchers built a breadboard decoder using about 100 integrated circuits for full capability (40 characters by 20 rows) British teletext.

1980. Interest in reducing decoder complexity, together with equipment orders from SSS, led to the introduction of the Virtext* teletext decoder. As new integrated circuitry became available, the decoder was reduced to some 35 silicon chips, a practical decoder design.

As an extension to the Virtext decoder later that year, Zenith developed the Virdata system, which allowed the SSS satellite-delivered services to feed additional cable TV systems.

1981. A separate Videotex Engineering Department was organized at Zenith as part of the Cable TV Products Division in January 1981. In its first year, the videotext research and development operation posted several significant developments:

- Design of a 25-inch/color TV sets with built-in teletext decoders,
- Development of prototype 13-inch viewdata terminals,
- Design of a special encoder for Telidon (the Canadian Standard) graphics, and,
- Preliminary research and development of a decoding system for full-field teletext based on the standard now known as North American Basic Teletext Specification (NABTS).

Decoder-equipped 25-inch diagonal color TV sets developed using British teletext technology were used in a teletext field trial by Field Electronic Publishing (now KEYCOM Electronic Publishing).

The Chicago-KEYCOM venture was the first commercial broadcast teletext experiment authorized by the FCC. Teletext receiving equipment was supplied to Taft Broadcasting Co., another teletext pioneer, for a similar teletext operation in Cincinnati.

The videotex engineering staff also designed a versatile teletext module that could be incorporated in either set-top or built-in teletext applications. This module included a remote control interface, allowing smooth integration into existing systems, and advanced display characteristics: 40 characters by 24 rows with 15 colors.

At the Western Cable Show in November 1981, Zenith introduced a special teletext decoder, called Z-TEXT, using this same module, Z-TEXT was developed for use with Zenith's tiered, addressable cable television system, Z-TAC.

Also in 1981, several customers expressed interest in 13-inch viewdata terminals. Prototype terminals, designed and built to Prestel specifications, offered full-color display and a detachable keyboard. Memory stored the user ID and terminal configuration characteristics. Several of these units were on display at Videotex '82 in New York.

1982. With the growing numbers of experimental teletext operations in 1982, Zenith supplied specially designed decoders to Time, Inc., for the field trials in San Diego, Calif., and Orlando, Fla. (Time ended its San Diego field trial in early 1984.)

The Time, Inc., operations offered full-field North American Basic Teletext Specification (NABTS) teletext. The special Zenith-designed decoders consisted of a Z-TAC cable TV decoder and a second module that includes a sophisticated teletext decoder system.

A broadcast quality teletext origination system, Virtext System 1, also was developed in 1982 to offer low-cost teletext origination. Based on a Zenith Data Systems desktop microcomputer, the editing and encoding system allows broadcasters to create, maintain and broadcast locally originated text service.

1983. Zenith joined forces with Taft Broadcasting to bring the nation's first commercial teletext service to consumers. Zenith entered into a five-year agreement with Taft in January to produce teletext receiving equipment for Taft's new commercial teletext service in Cincinnati.

The venture is the first U.S. teletext agreement between a television set manufacturer and a broadcaster. Under terms of the agreement, Zenith is producing teletext receiving equipment, based on World Systems Teletext (WST), the British standard, to meet market demand in the Cincinnati area.

In June, Zenith introduced its new full-field Teletext Decoder. Full-field teletext uses a full cable channel, providing access to up to 5,000 pages of world, national and local news; weather, sports, financial, and other fast-breaking information, as well as special features and services such as computerized banking and shopping.

1984. Zenith teletext equipment made an appearance at the 1984 Summer Olympics. Visitors to the Los Angeles area received up-to-date news and information on the games through the company's teletext decoders and color television receivers placed at the Olympic sites, hotels, airports and other high-traffic areas.

Field trials of the Zenith full-field WST system will begin this year. Current work at Zenith focuses on enhancements to existing products, including full-channel teletext.

Zenith is also bringing videotext, the communications technology of the future, into the present with a new personal computer (PC) and personal information terminal. The Z-150 PC—an IBM compatible personal computer introduced by Zenith Data Systems Corporation (ZDS)—and the low-cost ZTX-11 computer terminal already are being used by small businesses to retrieve videotext.

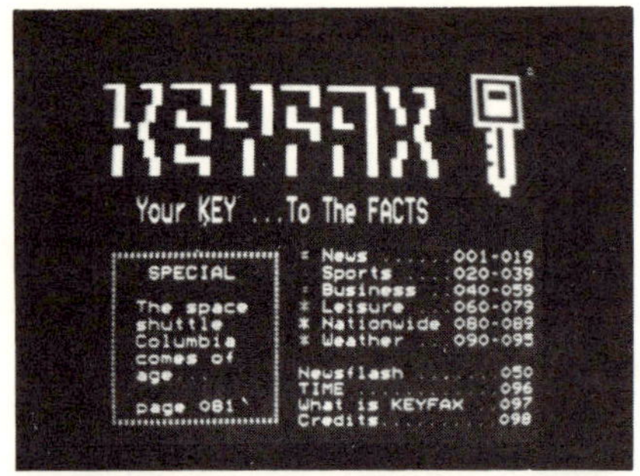

Figure 6.8 *Several of the indexes offered by Keyfax. Top left shows menu.*

NEWS

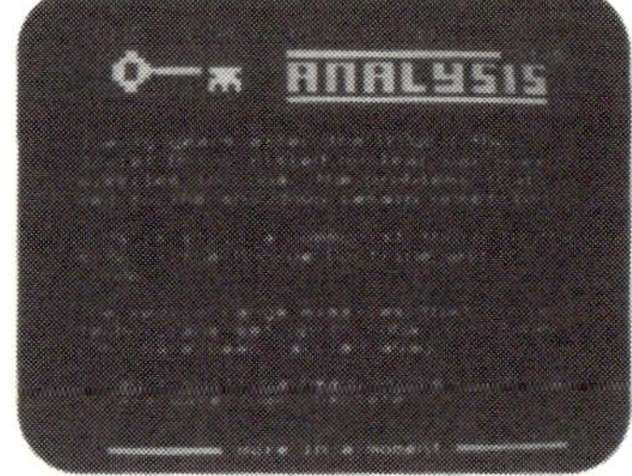

BUSINESS

SPORTS

Figure 6.9 *Features of Keyfax Teletex Service.*

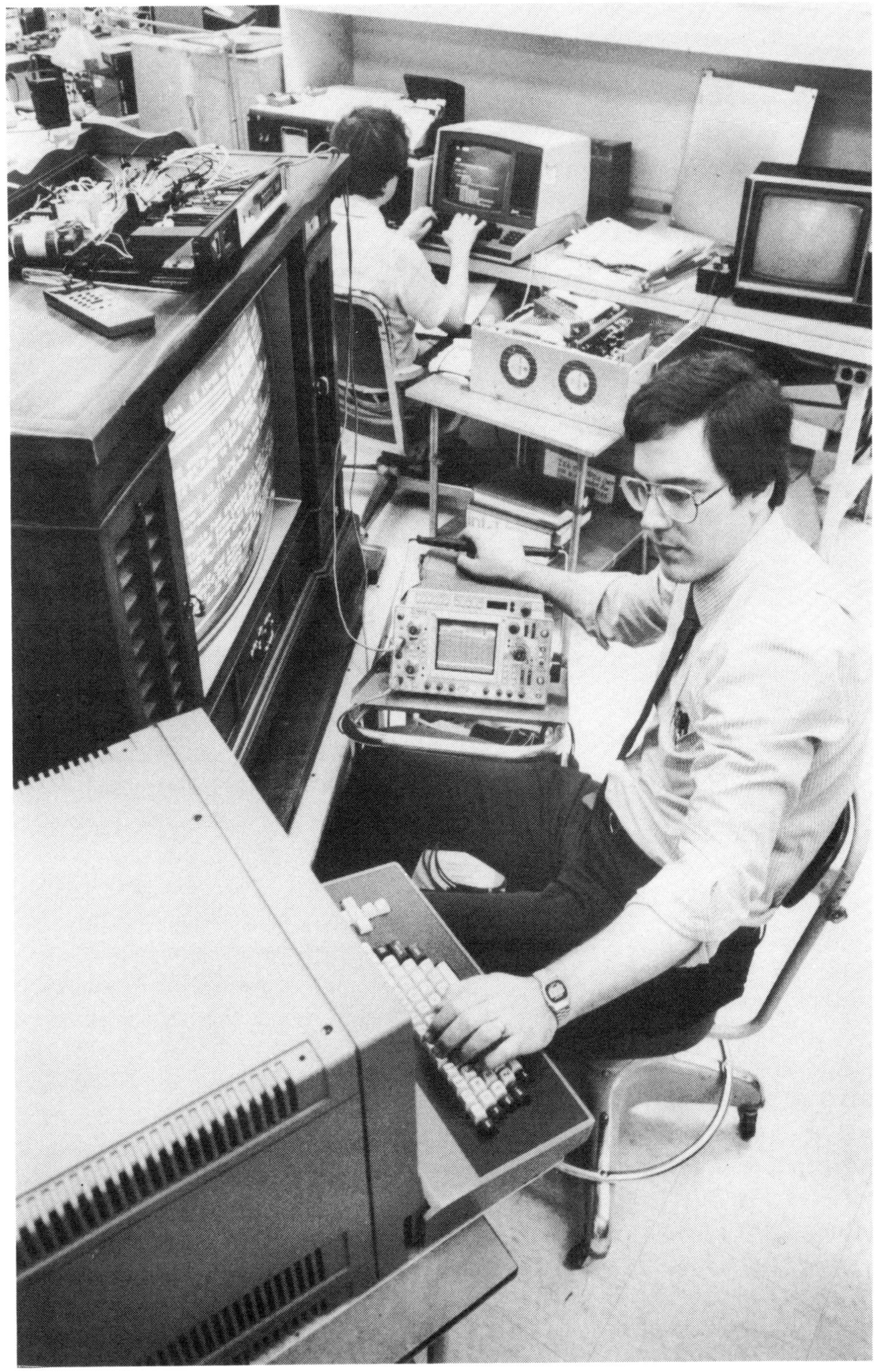

Figure 6.10 *Zenith Researchers have been actively pursuing research and development in videotex and teletext technology since 1978. Above, Gregory Woodsum, manager of videotex engineering, tests a World System Teletext decoding system at the company's Technical Center in Glenview, Ill.*

More significant are the broad videotext applications for these products in the future. Personal information terminals and personal computers have laid the groundwork for the burgeoning videotext industry. Thousands of Americans use videotext today—services such as Dow Jones News Service, the Source and AgriData.

The computer and the terminal both have the capability of accessing these established videotext services to receive the latest news, sports, weather and stock prices. With the appropriate software, either could be used for banking and shopping from home; in fact, almost anything in printed form eventually could be accessed—airline schedules, theater times, restaurant hours, even telephone listings.

The ZTX-11 has exciting videotext applications for the future because of its built-in modem and automatic dialer that can access not only computer information networks, but also business and educational mainframe computers, and, unlike many terminals available today, with the ZTX-11, users can access data bases with just one key stroke.

Likewise, the Z-150 is promising because its full-color graphics capability will allow it to display more sophisticated forms of videotext once the software becomes available.

Zenith's broad experience in high-quality displays, teletext and computers provides a solid foundation for growth in the emerging videotext market. The Z-150 and the ZTX-11 are part of the ZDS line of desktop and microcomputer systems, video monitors, terminals and software. See **Figure 6.11.**

The recent announcement by the Taft Broadcasting Company of Cincinnati, Ohio prompted a quick trip to the area to pick up a new Zenith decoder from the Tracey-Wells Company, the local Zenith distributor. After picking up my decoder, I visited several local Zenith dealers to get the full story and see this system in operation with Taft's, WKRC-TV station. All in all, I was impressed with Taft's new "Electra" service, the programming was good, the news, business and other offerings seemed worthwhile.

One final stop to a dealer on the way home was indeed an eye opener! This dealer was in the process of doing what I was to do in a few hours; he had a satellite set up and was attempting to pick up a teletext signal direct from Satcom F3R transponder 6 (WTBS). After a few minutes, he lead me to the service department to show me how he did the job.

The Zenith decoder was designed for use with the new Zenith Smart Component TV Systems, now on the market. While I liked the new Zenith system, I had no plans to purchase this system just to do the teletext bit, as I had just purchased a complete Sony system.

A quick look at the back of the decoder showed an input with "F" connector and an 8 pin DIN plug as an output to the TV set up. A 15 minute study of the schematics on the decoder, and the input of the Zenith system showed the output from the DIN plug to the TV (or monitor), to be the standard red, green and blue signal input. Another 15 minutes were spent making a few sketches of both units, plus the Sony input, which solved my upcoming problem—connecting the Zenith decoder to my Sony monitor. At this point, I went back to the dealers setup in the service department to learn the following:

(A) The Zenith decoder works great, as long as you have a clean sparkle-free transponder 6 (WTBS) signal. After this point, I found it necessary to fine tune the WTBS signal to get error free reception of the teletext service. With transponder 6, the weakest signal on F3R, some TVRO setups might have a problem with teletext reception.

(B) A 10 foot quality antenna would be the minimum size that would give error-free copy. The TVRO receiver should be stable. If not, you are going to be jumping up and down to tune the receiver every few minutes. I don't think some of the low end receivers will do the job, therefore try a decoder before you buy if possible.

(C) The Zenith decoder puts out red, green and blue signals to the video unit, therefore you must use a new monitor or one of the newer generation component TV systems to use this unit. It is possible to bring out the red, green and blue inputs on some TV's; your service dealer should be able to do this at a reasonable cost.

(D) Don't look for any audio on the teletext service, there isn't any. If you desire audio you could tune one of the many music programs found on the audio subcarriers of WTBS with a stereo processor.

The cost of the Zenith unit is under $300, and you should be able to purchase a unit from any Zenith dealer or distributor in the area that is now in service, Cincinnati, Ohio. The Zenith distributor in Cincinnati, Ohio is the Tracy-Wells Company, phone 513–621–1969, or Zenith Radio Corp., 1000 Milwaukee Avenue, Glenview, Illinois 60025, phone 412–391–8181.

Figure 6.11 *Zenith's terminal, Model ZTX-11, can be used to access established videotext services. This terminal will also access other data services.*

Courtesy Zenith Corp.

CHAPTER SEVEN

OTHER VBI SERVICES (TxT/VBI)

OTHER VBI (TxT/VBI) SERVICES

CableText is the umbrella name Satellite Syndicated Systems, Inc. has given to all of its satellite-delivered text services. Currently, CableText features a variety of news and information services, all of which keep subscribers up to date on the latest sports, business and general news events affecting our world. CableText utilizes the vertical blanking interval (VBI) of the WTBS, Atlanta, Georgia satellite signal to provide a national data distribution network. The WTBS signal is the most widely distributed satellite signal, servicing more than 6,600 cable systems in more than 29 million cable households.

While viewers need to be on a cable system to receive the WTBS satellite service, many of the potential users of CableText need not be on cable. Instead, the cable "headend" (where the satellite signals are received) can serve as a "gateway" through which nearby users can receive the necessary information and data via a phone line connection.

SSS CableText services are delivered in three modes:

Mode 1: The data is delivered directly to the cable TV subscriber's TV set, where the information is decoded and displayed on the TV screen (e.g. Keyfax National Teletext Magazine). Users need to be on cable for this mode, and this is generally the mode of delivery referred to as "teletext."

Mode 2: The data is delivered to a subscriber's computer or printer. In this mode, a specialty user has a decoder which attaches the cable system to a computer or printer; however, this type of service can generally be delivered over phone lines when a nearby cable system's satellite dish is used as a data gateway.

Mode 3: CableText began operations in this mode. The data is delivered to the cable headend, then the cable system creates a special TV channel for its subscribers using the data received. Usually this is a scrolling or carouseling series of alphanumeric "pages" that a subscriber views on his TV set when tuned to the channel chosen by the system for such a service. This mode is not limited to headend decoding only. The service may be decoded (provided the cable system carries WTBS) by using a normal cable TV hookup. A demodulator is needed to give a video output which is then fed into the teletext decoder. The output of this is fed into a TV monitor for viewing.

Within the WTBS satellite service is the vertical blanking interval (VBI) which was mentioned earlier. The VBI is the black horizontal line you may have seen at the top of your television picture. When the picture is properly adjusted, the VBI is not visible. Within the VBI, a one-way data stream can be carried. The use of the VBI was pioneered by the British more than 10 years ago as a means to handling data, and the first application was captioning for the hearing impaired.

There are approximately 20 lines in the VBI, six of which are currently used by CableText for delivery of data. Each line carries the equivalent of forty, 300 baud or ten, 1200 baud channels.

SSS has been involved in the development of VBI data delivery applications since 1978, with commercial start up of the CableText service in 1980. The transmission of UPI and Reuters cable news services inaugurated the commercial service, and today, those services in addition to Dow Jones and Associated Press are delivered to cable headends. Also, two real-time commodity services are carried via satellite to specialty users. This is an application of the mode 2 delivery, wherein the commodity information is "downloaded" to a computer setup at the user's location. A CableText decoder feeds the data into the computer via an RS-232 standard cable interface. The computer is pre-equipped with software to analyze the commodity information, create graphs and charts and initiate buy/sell orders and recommendations.

Since May 1982, CableText has been carrying the Keyfax National Teletext Magazine, which is an example of the mode 1 application. During latter 1983, participating cable subscribers will have a decoder box in their home to individually decode and display preselected news and other current information on their TV screens.

CableText is capable of delivering any kind of predetermined, categorized data or alphanumeric information to any point in the contiguous United States, Alaska, Hawaii, Puerto Rico and the U.S. Virgin Islands. Since the VBI of the WTBS satellite service is used, the services carried in the VBI may be continuously cycled. The WTBS signal is offered 24 hours each day without any interruptions for local commercial insertion.

At this time, it is not clear as to SSS position on the sale and use of CableText decoders by individuals and TVRO owners.

PROGRAM INFORMATION NETWORK (PIN)

PIN, the acronym for Program Information Network, is a new SSS electronic mail service, which will enable cable networks to deliver urgent information and extremely current programming data to their cable affiliates. Future uses may also include MSO electronic mail, newsletter distribution and other one-point to multi-point functions, which would serve as an industry-wide bulletin board.

SSS CableText provides common carrier service on the Vertical Blanking Interval (VBI) of transponder 6, Satcom 3R satellite signal (WTBS). The CableText Division coordinates and markets the data space in the VBI, including PIN.

The monthly fee of $500 is charged to the "senders of information" This fee includes the transmission of one original message each business day, for a total of up to 20 messages a month. Each message will include up to 35K of data, or approximately 5 to 6 pages of text. The receivers of information will not be charged to receive this service, however, they will have to purchase the necessary hardware to receive the messages. The cost of receiving equipment ranges from $800 to $1,200.

The "senders of information" will create an ASCII text file, which they will deliver by telephone to the SSS satellite uplink facility in Douglasville, Georgia. Then, on a first-come, first-serve basis, the data files will be transmitted into the VBI of the WTBS satellite service. From there, the data file will be encoded and addressed to all affiliates of senders who are equipped with a teletext decoder and printer. At the receiving end, the data will be decoded and automatically printed in "hard copy" form for easy reading.

Duplicate transmissions of each message will be sent throughout the day. If a message has already been received, the decoder will be programmed to ignore the duplicate transmission. If the message was not received the first time it was sent, it will print out during one of the duplicate transmissions. **Figure 7.0.** The following equipment is needed.

Senders:
1. A computer equipped with a communications port.
2. A word processing program to create an ASCII text file.
3. A 300 or 1200 baud modem to tie into the computer located at the SSS uplink facility.

Receivers:
1. The receiving sites will need to be receiving the Satcom 3–R, transponder 6 service (WTBS). If you already receive WTBS, the video output from the receiver or a cable connection can be used to receive the VBI data.
2. A decoder and printer to decode the VBI data, and then print it out in "hard copy" form for easy reading.
3. If decoding is done off a cable drop (RF feed), a demodulator is required to provide a video feed for the decoder. The demodulator can be obtained through Zenith, the decoders from Texscan/ Compuvid or Zenith and printers from STAR or Epson.

DOW–JONES NEWS SERVICE

This text-type service is carried with the VBI on Satcom 3R, transponder 6 (WTBS) via the SSS system, and is a primary business service used by cable systems. However, this service may be subscribed to by an individual TVRO owner by contacting the company. The basic monthly charge is $150 for all information carried. **Figure 7.1.**

The Dow-Jones (Wall Street Journal) System uses a VDS DOW 600 Character Generator and a Zenith RS232 Decoder Unit. This equipment can be purchased outright or leased by the individual for approximately $65 per month, for each of the above units. The service features the following on a daily basis:

STOCK MARKET—

Overseas Report, plus Pre-Opening "Outlook," capsulized versions of columns "Heard on the Street" and Abreast of the Market. After market close — 10 Most Active Composite Trades on NYSE and AMEX — Hourly Most Active Stocks on NYSE and AMEX — Closing Prices 300 Most Active — NASDAQ Closing — Market Diaries — NYSE, AMEX and NASDAQ — Consumer Interest Rates — Lipper Municipal Bond Fund Index — Complete NYSE Exchange Prices — Capital Markets — Commodity Reports and Future Markets — Gold and Silver Markets — Dollar Market — Business News — plus an extensive weekend report with many features from the Wall Street Journal and Barron's columns and features on a selected basis.

ASSOCIATED PRESS NEWSCABLE

AP Cable News is on the SSS VBI section on Satcom 3R transponder 6. Features 24 hours per day, seven days a week of national and international news, sports and business reports from the Associated Press Bureaus worldwide.

This service is text in nature, and is a cable pay service, requires special VBI Decoder Unit, not now available to individual subscriber. **Figure 7.2.**

UPI CABLE NEWS

An alphanumeric text service from the 75 year old United Press International Services. This text programming was set up for the broadcaster, CATV, SMATV, MDS, and other subscriber related services as a 24 hour, news programming service that could be used to generate local advertising income.

UPI is not a newcomer to satellite delivery of its news and radio news programming. In 1962, UPI started dispatching its news service to a few subscribers on a limited basis. Toward the middle of 1976, UPI was engaged in full testing of an overall distribution system via satellite.

UPI cable news uses the Vertical Blanking Interval (VBI) of WTBS on Satcom 3R transponder 6, and runs 24 hours a day. The VBI decoder is a Zenith RS232 output located at the cable or broadcaster's satellite receiver facilities.

This service is available only to cable systems and requires special decoding VBI teletext receiving equipment. **Figure 7.3.**

"UNIQUE" SATELLITE SERVICES
GENESIS

In the unique category is a service known as Genesis Storytime. This service is transmitted on F3R TR8 (CBN), but it is not transmitted in the VBI, but instead by means of the subcarrier. A modulated tone is transmitted. To decode the signal, one uses what is essentially a "videotex" terminal. Instead of hooking the videotex terminal to the telephone lines, it is hooked up to the audio output of the subcarrier demodulator. The operating system is *NAPLPS,* which is the videotex version of NABTS. In layman's terms, they are using the North American videotex system rather than the French or British system. There is a significant cost saving involved, and the terminal equipment is more widely available from such sources as A.T.& T., Electrohome of Canada, Norpak of Canada, AEL Microtel of Canada (a G.T.&E. subsidiary) and soon, *SONY.* Terminals cost roughly $1,000 for the Genesis Storytime system. It *might* be possible to get pictures on an IBM PC, using some special software. Genesis transmits children's stories with graphics and texts. Cable operators are picking up the text using Wegener audio demodulators, and they take the decoded videotex pictures and remodulate them on a full channel of their cable systems. *SSS* is handling the marketing in the U.S. of the service. Thus, using a narrow bandwidth high resolution, TV graphics are transmitted. **Figure 7.4.**

"IN-TOUCH"

A new audio service that provides a free reading service for the blind, and other print-impaired persons. Everyday, volunteers read text from well-known periodicals as the Wall Street Journal, Barron's, Sports Illustrated, Good Housekeeping, Vogue, and many other publications. "In-Touch" is transmitted from Satcom 4, transponder 3.

Southern Satellite Systems furnishes this as a free, no-charge service to the cable companies as a good corporate citizen and works with the "In-Touch" group. This group is a non-profit organization run by volunteers; it broadcasts information useful in everyday life. Besides in-depth reports and magazine

UP LINK HEADEND
DOWN LINK HEADEND
CABLE SYSTEM LINE
ENCODER
DECODER
SSS CPU
AFFILIATE PRINTER
AFFILIATE COMPUTER
ALTERNATIVE FEED IS OVER PHONE LINE
PHONE LINE MODEM TO MODEM
SENDER COMPUTER
AFFILIATE PRINTER
AFFILIATE COMPUTER

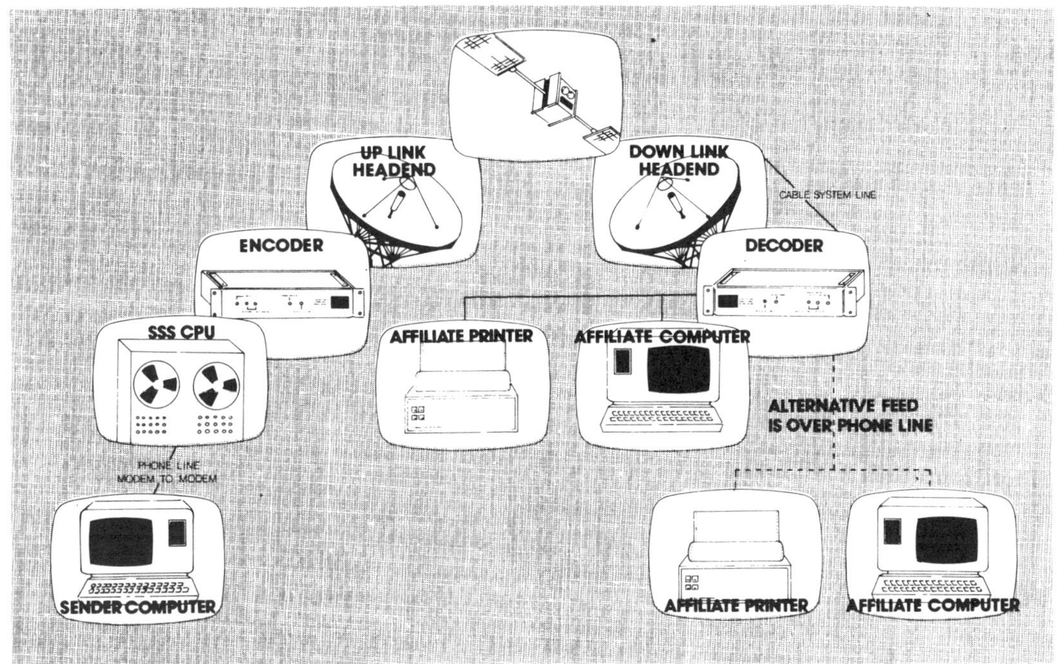

Figure 7.0 *The Program Information Network (PIN) system.*
Courtesy Satellite Syndicated Systems, Inc.

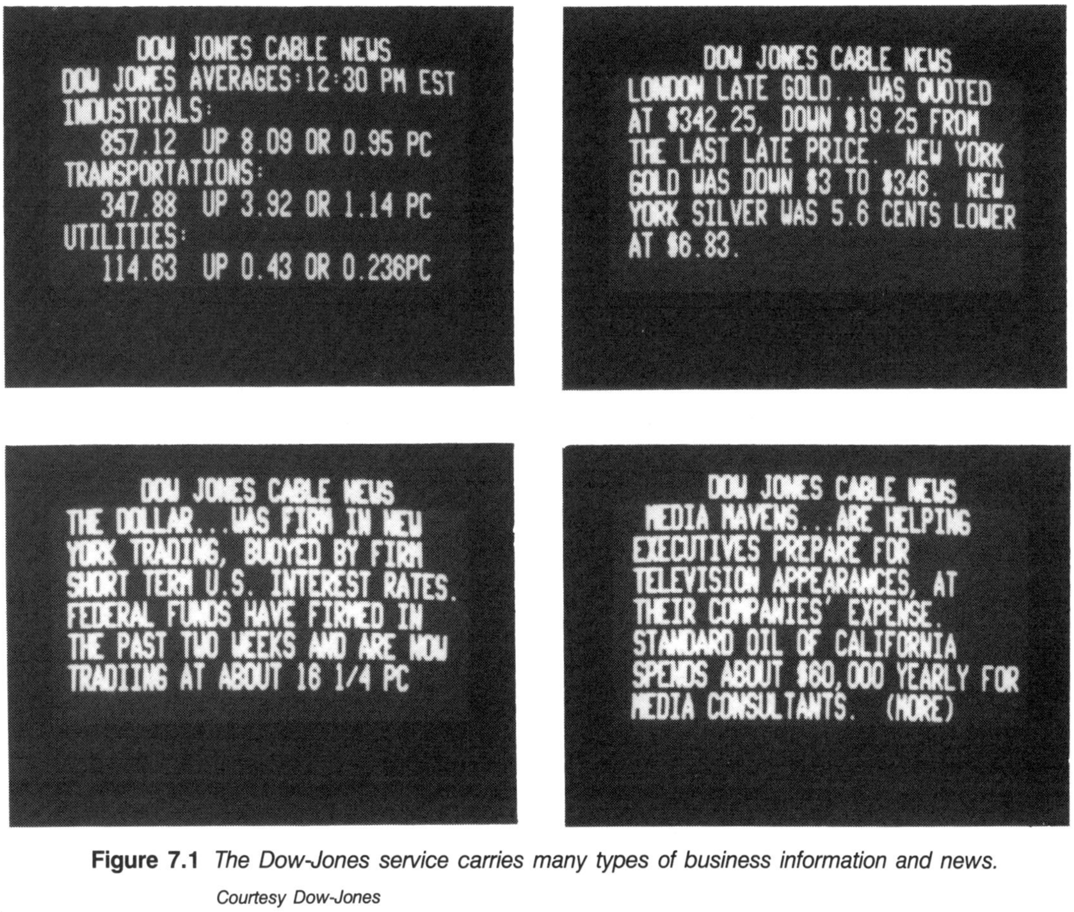

DOW JONES CABLE NEWS
DOW JONES AVERAGES: 12:30 PM EST
INDUSTRIALS:
857.12 UP 8.09 OR 0.95 PC
TRANSPORTATIONS:
347.88 UP 3.92 OR 1.14 PC
UTILITIES:
114.63 UP 0.43 OR 0.236PC
DOW JONES CABLE NEWS
LONDON LATE GOLD...WAS QUOTED AT $342.25, DOWN $19.25 FROM THE LAST LATE PRICE. NEW YORK GOLD WAS DOWN $3 TO $346. NEW YORK SILVER WAS 5.6 CENTS LOWER AT $6.83.
DOW JONES CABLE NEWS
THE DOLLAR...WAS FIRM IN NEW YORK TRADING, BUOYED BY FIRM SHORT TERM U.S. INTEREST RATES. FEDERAL FUNDS HAVE FIRMED IN THE PAST TWO WEEKS AND ARE NOW TRADIING AT ABOUT 16 1/4 PC
DOW JONES CABLE NEWS
MEDIA MAVENS...ARE HELPING EXECUTIVES PREPARE FOR TELEVISION APPEARANCES, AT THEIR COMPANIES' EXPENSE. STANDARD OIL OF CALIFORNIA SPENDS ABOUT $60,000 YEARLY FOR MEDIA CONSULTANTS. (MORE)

Figure 7.1 *The Dow-Jones service carries many types of business information and news.*
Courtesy Dow-Jones

Figure 7.2 *Sample of screen of AP Cable News Service*

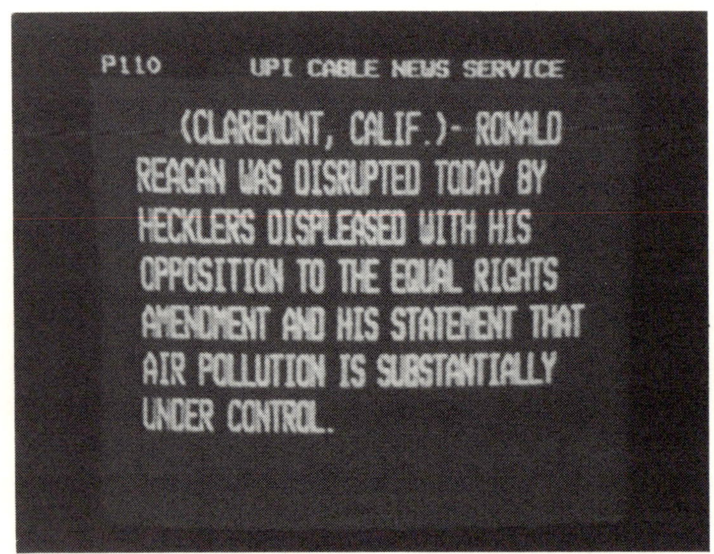

Figure 7.3 *Screen sample of the UPI Cable News Service*

readings, they offer supermarket sales, stock quotations 24 hours a weekday, and 14 to 15 hours a day on weekends. This service represents a $125,000 subcarrier donation by Southern Satellite Systems, and they should be commended for furnishing this much-needed service to the print-impaired citizens whose need is filled by this generous donation.

The service is contained in the Standard Subcarrier System and can be received by standard receivers fitted with a tunable subcarrier feature. Again, the service is on a no-charge basis.

BEETS DATA SYSTEMS

Located at a Satcom 3R, transponder 6, Vertical Blanking Interval is the BEETS Data System, which is a basic commodity trading service that provides real-time commodity quotes and financial information for downloading into a home computer setup. It furnishes charting for 20 commodity classes, lists 100 contracts on 10 pages of text. This system is menu-driven and a subscriber can choose material from listings.

An Apple Computer is the basic hardware, with the software furnished by the BEETS organization. An additional decoder box is required, along with a serial board if required to drive your serial printer. This is a basic, 300 baud rate service, furnished through the CableText services for decoding of the commodity information at subscriber's home or office in conjunction with a computer and/or printer.

BONNEVILLE DATA SYSTEM

A new and very novel means of data transmission and complete commodities coverage is offered by the Bonneville Telecommunications Data Systems Division. This system is known as Market Monitor®, and uses Bonneville satellite and FM station networks to relay the data to the subscribers.

This series relay starts with an uplink of the commodities data to the satellite, downlinked to a local FM radio station in the subscriber's area. The subscriber has a decoding box, that attaches to an FM radio to decode data on the subcarrier in the FM signal. This network utilizes many FM radio stations by using the capability of their audio subcarrier. After receipt, the data is then routed through an IBM personal computer. Naturally, the IBM PC requires the use of their proprietary software in order to select and read out the desired data.

The Bonneville System gives complete commodities coverage and provides true tic-by-tic coverage of the following markets: Precious Metals, Financial Instruments, Agricultural Commodities, Cash Markets, and Petroleum.

The software package known as OCTOOL allows you to track the time of every tic, make graphs, set stops on over 80 commodities, and create a total history for each of these commodities. This service is widely used by many businesses and individuals who daily trade in the commodities market. This is not a VBI service. For full information, call Bonneville Telecommunications Corporation, Data Systems Division, 1–800–453–9404.

Figure 7.4 *Computer-generated children's storybook channel that displays full color pictures and words on the screen page-by-page.*

CHAPTER EIGHT

MISCELLANEOUS SATELLITE SERVICES

REUTERS SERVICES

Reuters News services spans a time frame of over 125 years of continuous service to the public and business world. This highly respected name in the news-gathering world is also one of the pioneer users of the satellite systems, and other state-of-the-art techniques in the modern communications world.

Reuters originally set up their monitor service for Wall Street brokers and financial houses. After years of successful service, the Reuters Monitor Service was offered to the general business world via satellite.

The service is transmitted off Satcom F3R, transponder 18 during the business day. Many think this signal and data is a scrambled video picture: it is not, it is a system of high speed data transmission developed by IDR, Inc., of Farmingdale, New York, a wholly-owned subsidiary company of Reuters.

This system uses a row-grabbing technique, which converts the data and or test into packets; from this point to a digital stream to a standard television format for broadcast to the subscriber's terminals. The subscriber's terminals can be hooked up to a cable system, or to an over-the-air satellite receive-only system. The Reuters Service requires a special terminal with the proper software in order to receive the data and display the data on the terminal's CRT, or on hard copy with the use of a dot matrix printer. The unit's terminals are addressable and can be activated only by the proper signals, which are sent along with the data stream to turn on and turn off service to a defined terminal.

The Reuters System carries many forms of commodity, stock price quotations, money market prices, foreign exchange, plus a wide range of general interest and business news. The nature of the data carried requires the Reuters System to be accurate and free of errors. Because of this high quality, the system is constantly checked for low-error rate.

The Reuters signal can be transmitted as an FM double sideband signal or at video baseband. Reuters terminals will display up to 64 characters on 16 lines, and are capable of carrying up to 1,000 pages in the system. **Figure 8.1.**

The Reuters Monitor Service covers the following information groups with many pages of information:

Money/Financial Futures
Grain/Livestock
Metals
Soft Commodities
Coins
Energy
Securities
News/Business
International News
Contributed Information
plus other information

Private individuals may lease the Reuters terminals and use their present TVRO systems to receive the data off the air. Contact Reuters, 2 Wall Street, New York, New York, 10005, 212–732–7800.

REUTERS SMALL DISH SERVICE (SDS)

The Reuter Monitor Small Dish Service (SDS) combines proprietary state-of-the-art computer technology and advanced communications systems with Reuters renowned news and information-gathering superiority.

Utilizing the most advanced direct satellite transmission techniques, the REUTERS SDS SERVICE offers a highly affordable professional terminal that can be tailored to individual needs.

The SDS SERVICE offers up-to-the-second price quotations on almost all U.S. commodities and many foreign contracts as well. In addition, recipients will have access to the latest news and information affecting the markets, as well as trading summaries and commentaries.

Detailed coverage of U.S. money markets and rates, foreign exchange quotations and reports, stock market reports and Federal Reserve Bank activity is also available on the service.

At the heart of the system is a series of unique "personal pages," which allow the user to custom design screen displays to instantly provide the most desired information from the entire spectrum of available news and data.

The SDS SYSTEM utilizes a new generation REUTER-developed microprocessor and software, designed to exploit the field of broadband satellite communications.

Broadband transmission makes it possible to access the Reuters satellite signal with a small receiving dish, which in the case of the SDS service, is a two-foot dish that can be easily mounted in any convenient location.

The small dish receiver allows the user to obtain the REUTERS signal 24 hours a day, anywhere within the satellite service area, subject to line of sight. **Figure 8.2.**

Reuters small dish service covers the following information areas: Grains, Weather, Livestock, Business/Market, Foreign Exchange, Money, Metals, Energy, Sports, Gold, News, Stocks, plus others. The service carries many pages of information in each of the above categories.

A teleprinter service gives access to the highly regarded report areas of information. These are: The Reuter Money Report, The Reuter Metals Report, The Reuter Commodity Report, The Reuter Grain and Livestock Report, and The Reuter Financial Report. This small dish service is an excellent service and desirable to the individual user because of the small dish size (24 inches), yet gives all of the needed services.

This SDS Service is very complex in nature and is totally addressable by the company (turn "on" or turn "off"). The signal is contained in the same transponder as the main Reuters Monitor (text retrieval) Service. The service is available to individual users on a reasonable monthly rate. Contact Reuters at 212–730–2715.

REUTERS NEWS-VIEW SERVICE

Yet another information and news service from Reuters is their News-View Alphanumeric Text Service. This service has been in constant use for over 10 years, and was originally a service for CATV operators.

News-View has been delivered by several methods—telephone lines and by Vertical Blanking Interval (VBI) on Satcom 3R transponder 6.

The service requires special VBI equipment to decode the signal, and is not available to individuals on a monthly basis, as the service is mainly a cable system news channel.

The service carries two types of information and programming. During the day its content is business and financial in nature, with some news content; in the evening hours, the service goes to a sport and sports scores reporting programming.

News-View carries a 10 to 12 minute cycle of contents, so it is possible to gather an up-to-date news summary in a short time. A feature of the financial channel of News-View gives the real-time stock prices of the most active stocks on the New York Stock Exchange, as well as key stock indices, company news, and key commodity news and prices.

COMMODITY COMMUNICATIONS NETWORK

On Satcom 3R, transponder 16, you can tune the audio subcarrier to 6.2 MHz and hear the digital audio signal of the Commodity Communications Corporation Network. This signal sounds much like buzzing bees or insects.

The CCC system utilizes several of the personal computers, Apple and IBM PC. The PC hardware and software service is called FutureSource™. The features of each computer system are shown in

Table I.

SEE NEXT PAGE

The CCC system also has three other levels of services for users; these are the VQP-3000 system, the VQP-5000 system, VQP World News, and Compuquote.

THE VQP-3000 SYSTEM –The most economical quote system, the VQP-3000 brings real-time prices, point and figure charts and commodity news. The keyboard is user friendly as function keys are marked for quick

FutureSource

FEATURES	IBM PC	APPLE
Charts	23 Rescaleable Bar Charts	16 Rescaleable Bar Charts
Programmable Pages	10 Programmable Pages. Each Page Tracks 14 Contracts	10 Programmable Pages Each Page Tracks 14 Contracts
Market Minder™	Sets High & Low Price Limits on 14 Contracts	Sets High & Low Price Limits on 32 Contracts
News	Scrolling Video News with Commodity World News	Interface to World News
Last 5 Trades	Display Last 5 Trades of Preselected Contract, Time of Day — Last Trade	
Split Screen	View 4 Features at Same Time	
Spreads	Display 4 Contract Spreads at Same Time	
Group Format Displays	Displays Last Price, Net Change for Seven Contracts for Grain, Meats, Stocks, Etc.	
Equipment Requirements	For IBM PC 128K, Also IBM Compatible Computers	Available for Apple II, IIe 48K or 64K — RAM

TABLE I

access to information as needed. The monthly fee includes the appropriate mode of transmission for fast and efficient data delivery anywhere in the U.S.A. and Canada. These include the Associated Press Satnet System, designed for the futures trader. **Figure 8.3.**

One of the features of this simple-to-use system is its single strike key system, which brings up the asked for information at once. The complete system covers: 7 Fixed Format Quote Pages on Grain, Meat, Metal, Money, Food Items, Financial 1 and 2, plus Petroleum.

Three additional programmable pages allow for formating 58 contracts with all changes, trend indicators, high/low, and bid/offer on any programmed contracts.

An all options page is furnished for every commodity, showing last change, high/low, bid/offer for all option months. Some of the other extras are two point and figure charts, which allow you to set scale, midpoint/and reversal for the commodities of your choice. These charts are updated automatically.

The five page commodity news pages keep you abreast of current fundamental information affecting commodity prices. **Figure 8.4.**

THE VQP-5000 SYSTEM – This system is a complete quote system for the professional trader who needs technical and fundamental information, in addition to real-time quotes. This quote processor was designed to assist in monitoring the market with efficient quote displays, rescaleable bar and point and figure charts and a price alert system. Programmable pages let the user tailor the system to meet his individual trading requirements. The program can be interfaced to computers.

The VQP-5000 carries the following services and features: 12 Fixed Format Quote Pages, 10 Programmable Pages, All Options Page, 10 Rescaleable Chart Pages, Market Minder™, Spread Page, and five Commodity News Pages. **Figure 8.5.**

For a small charge, VQP users can subscribe to a comprehensive news service, Commodity World News Service (CWN), which covers events pertinent

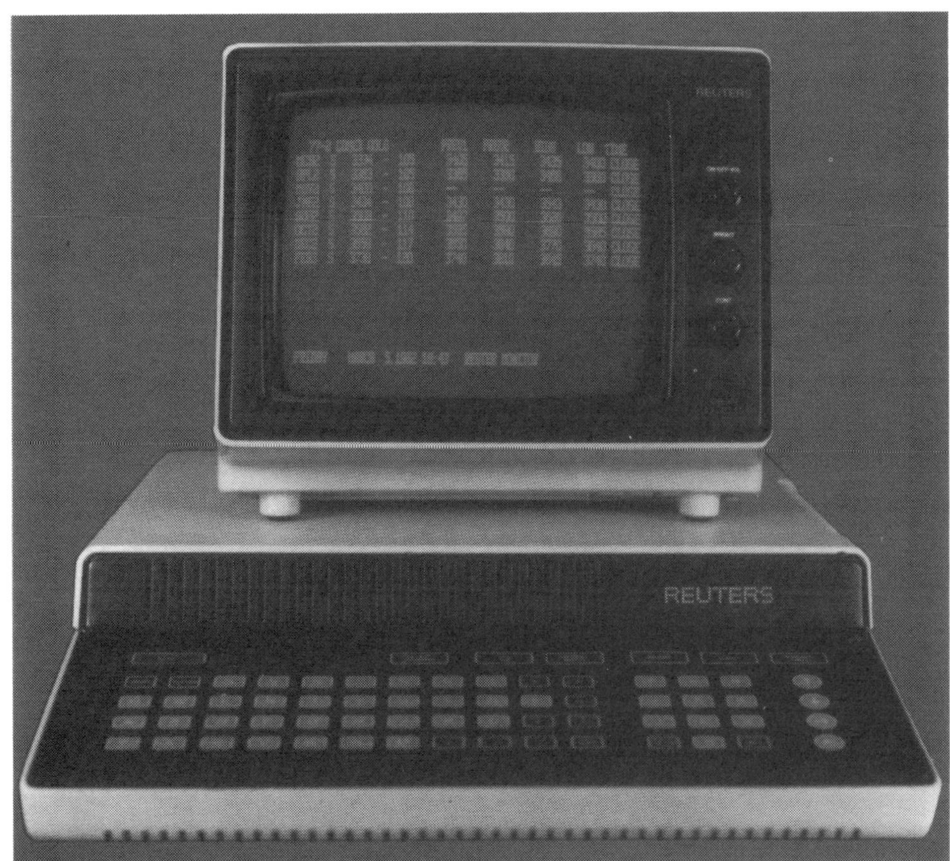

Figure 8.1 *Reuters Professional Data Terminal*
Courtesy Reuters

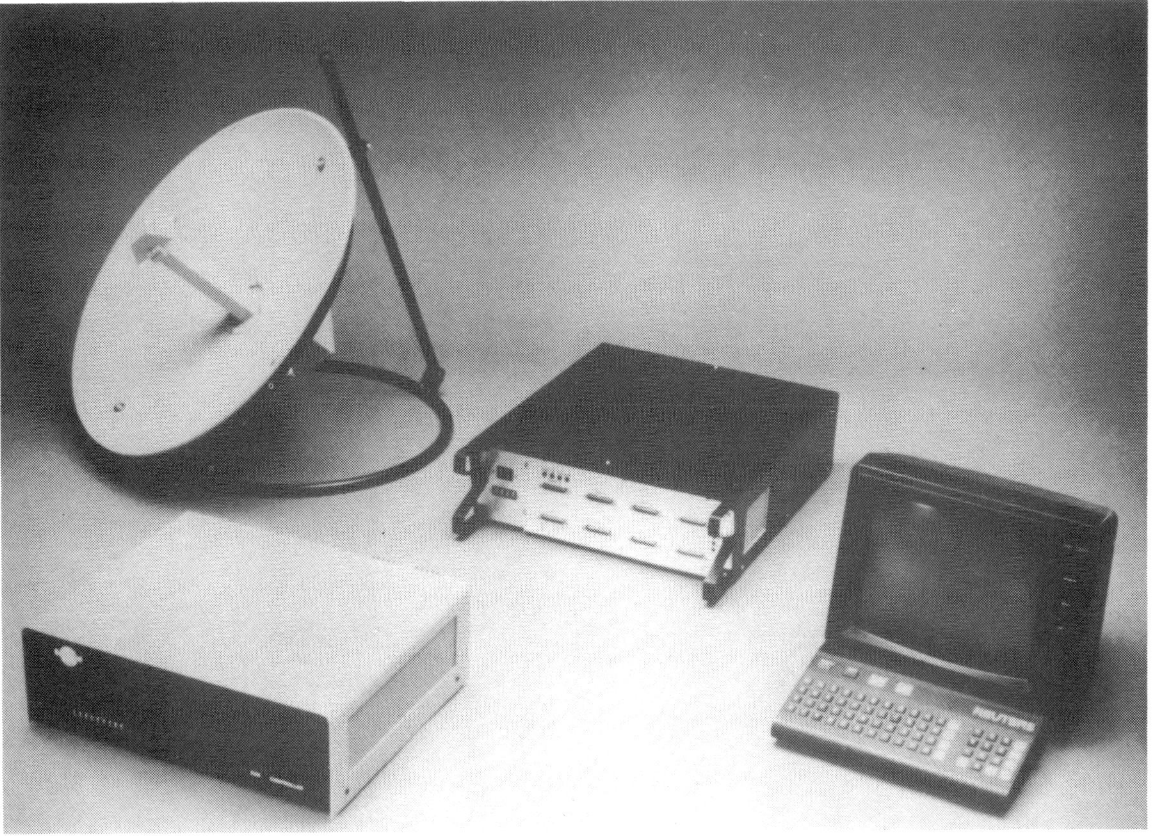

Figure 8.2 *Reuters complete small dish service equipment package with 2 foot antenna and mount.*
Courtesy Reuters

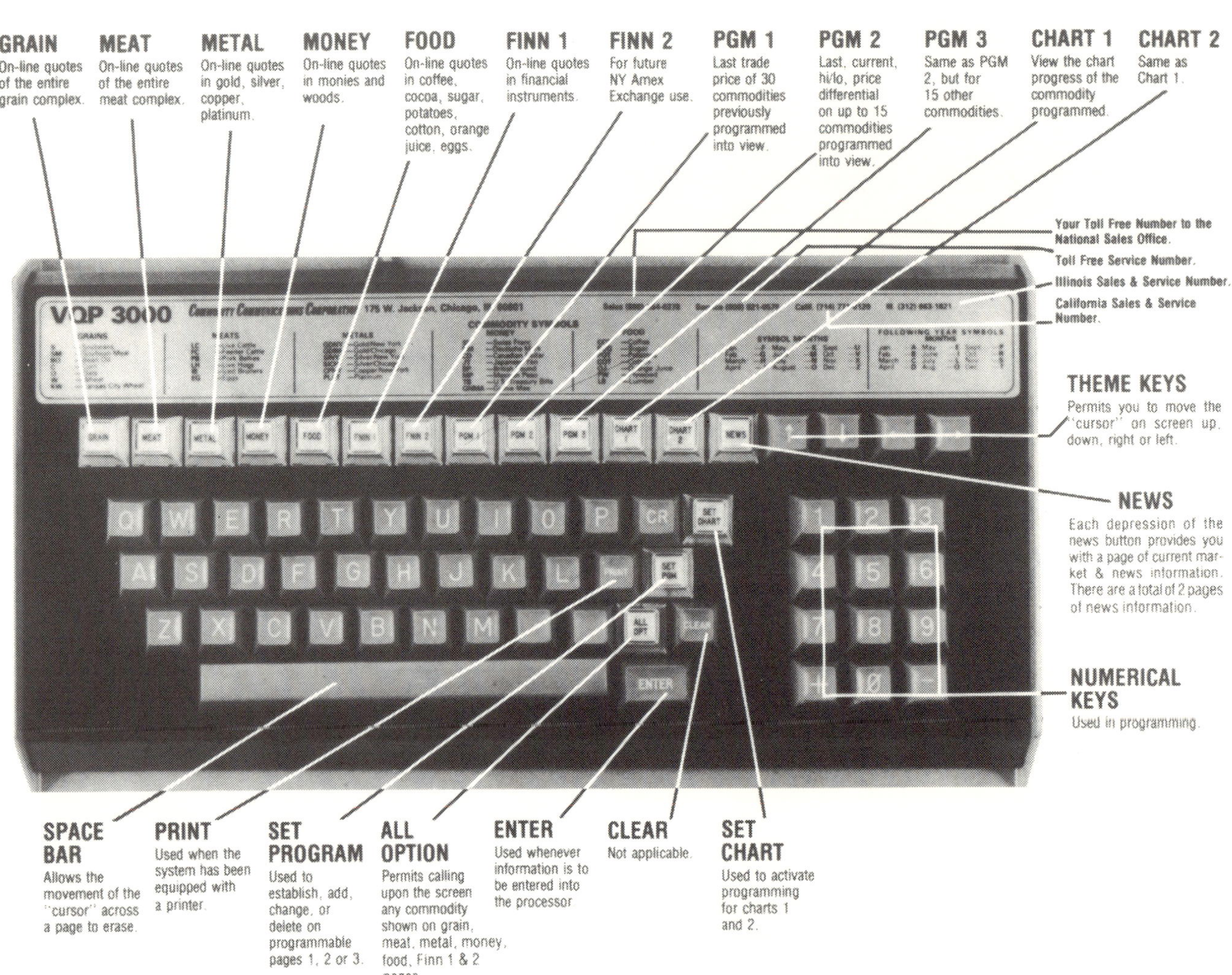

Figure 8.3 *The CCC VQP-3000 terminal showing all function keys and listings.*

Courtesy Commodity Communications, Corp.

```
SETTLEMENT PRICES                                     15:55
   TBQT    MCD      ED     USBD   GNMA    TYN      PA
U  9062U  9004U   8955U  7109   6820   7930 U 14785
    +13    +10     +13     +3    -23     +4     -305
Z  9025Z  8956Z   8912Z  7022   6715   7904 Z 14925
    +12    +11     +12     +2    -23     +6     -305
H  9002H  8925H   8883H  7006   6619   7815 H 15095
    +16    +13     +15     +3    -23     +7     -305
M  8985M  8903M   8863M  6925   6600   7731 M 15265
    +16    +15     +15     +3    -23     +7     -305
P  8970P      P       P  6914   6517   7717
    +17                    +3    -23     +7
T  8957T              T  6905   6504
    +18                    +4    -23
C  8944           8798C  6829   6425
    +19                    +4    -23
```

7 fixed format quote pages arranged by complex which display real-time prices. Screen shows last and change from previous close. Trend indicator appears when market reaches new high or new low for the day.

```
            SELECTABLE  PAGE
 COMM   PRICE    NET   HIGH   LOW    BID    ASK
GDNYQ    4199     -1   4235   4193
GDNYV    4250     -5   4288   4245
GDNYZ    4322     -6   4363   4315
GDNYG    4400     -7   4445   4395

SINYQ   12328    -52  12440  12360
SINYU   12370    -60  12660  12250
SINYZ   12710    -47  13010  12630
SINYH   13055    -35  13370  13040

CPNYQ    7350    -75   7350   7350
CPNYU    7385    -80   7495   7325
CPNYZ    7615    -85   7720   7560
CPNYF    7695    -85   7790   7790
```

3 programmable pages allow you to format 58 contracts you monitor most displaying last, change from previous close, and trend indicator, high/low, and bid/offer appear for any contracts you program. Strike the key once more to see opening/closing ranges, volume/open interest. Strike the key again for previous settlement and tick count.

```
  NEW YORK ESTIMATED SALES AS OF 10:00 CDT
GOLD                 11000  HEATING OIL            1380
SILVER                9000  POTATO/POT CASH       34/43
COPPER                4000  COCOA                  2321
PALLADIUM              330  COFFEE                  791
REG LEADED GAS         340  SUGAR 11/12        3588/482
PLATINUM              1670  CRUDE OIL               280

/CASH MONEY & BOND MARKET RATES AS OF 9:57 CDT/
FED FUNDS: 9 3/8 PCT      PRIME RATE: 11 PCT
11 7/8 PCT 10 YEAR NOTE: 101 01-03          UP .01
12 PCT 30 YEAR USBD: 102 16-18              UP .04
US 3 MTH T-BILL: 9.35 BID - 9.34 OFFER
13 PCT GNMA CASH QUOTE AUG/SEP: NOT AVAILABLE


 AUGUST 17, 1983                        WEDNESDAY
LONDON PM GOLD FIX: 422.50  *AM FIX: 420.25
10:21  ADV: 681    DECL: 563    UNCH: 406
10:29  NYSE VOLUME: 27,440,000
```

5 commodity news pages keep you abreast of current fundamental information affecting commodity prices.

Figure 8.4 *Several screen prints of the services on the VQP-3000 system.*
Courtesy Commodity Communications, Corp.

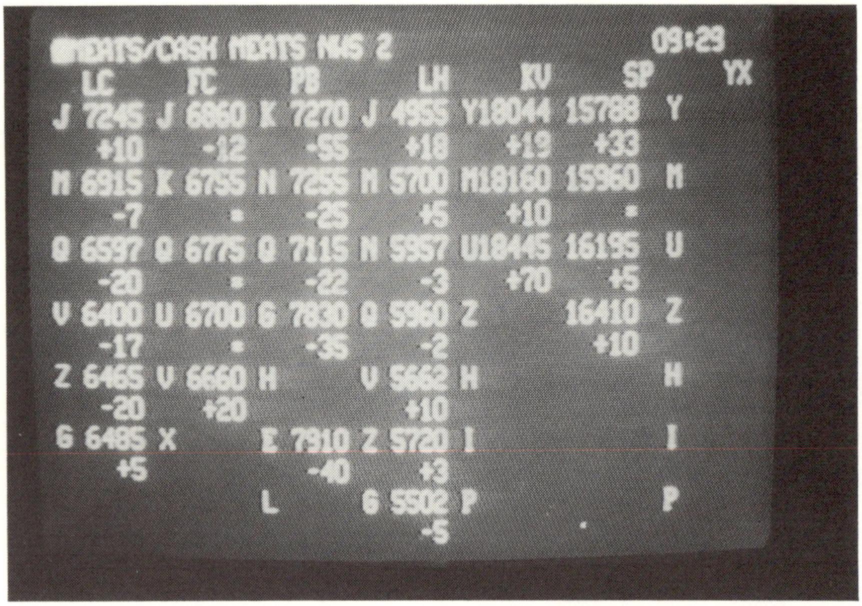

Figure 8.5 *Photo from screen of VQP-5000 terminal.*

to the futures markets more extensively than what is covered on the regular VQR System. Many editors, experienced in handling commodity information, gather and report the news as it happens. CWN sends in-depth reports on government and private sectors affecting the markets, weather, grain and livestock movement, floor trading from all major exchanges, financial auctions, and other sources. A printer must be used to receive Commodity World News.

COMPUQUOTE – This is yet another CCC service offered to the commodity traders and brokers. In short, it is a combination (single) commodity quotation and analysis system. CCC calls this an intra-day analyst. It is a program which tracks intra-day price movement of commodity futures contracts, and concurrently performs technical analysis on this accumulated data.

The intra-day system operates on an Apple computer interfaced to a Commodity Communications Corporation VQP-3000 or VQP-5000 quote machine.

A summary of some of the Compuquote configuration is: Oscillator Program, Spread Program, Ratio Program, Momentum Program, Relative Strength, Stochastic Program, Point and Figure and Moving Averages.

CCC will furnish a 2 foot or 5 foot dish antenna and monitors for subscribers use. **Figure 8.6.**

All CCC services can be printed with most standard dot matrix printers such as the Epson line with graphic features. The printers are furnished by the user or from Commodity Communications Corporation. There is now another fast and inexpensive way to reproduce the video information shown on the monitor screen. This new video printer from Mitsubishi is known as the P-50U Video Printer. This new video printer will print a total screen in about 15 seconds. The unit retails for approximately $500 plus. The P-50U will also produce photo quality prints about the size of a polaroid camera print. **Figure 8.7.**

In summary, the Commodity Communications Corporation seems to have the most complete overall systems which can be purchased and used by individuals. These systems can be purchased at any level of needed information. **Figure 8.8.**

THE GALAXY DATA SYSTEM — PRESENT AND FUTURE

The successful launch of Galaxy 1 in June 1983 by Hughes Communications was the first in the three-spacecraft Galaxy system.

Galaxy 1 is primarily a commercial cable satellite

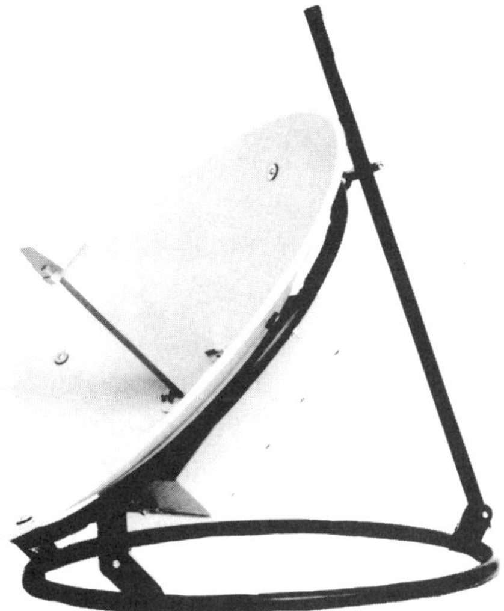

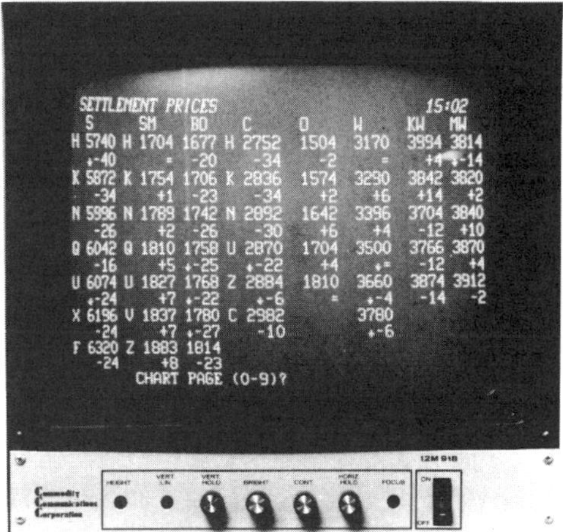

Figure 8.6 *Commodity Communications Corporation's 2 foot dish antenna and monitor.*

Figure 8.7 *New Mitsubishi P-50U Video Printer with copy from the VQR-5000 system.*

located at 134° west longitude, followed by Galaxy 2 located at 74° west longitude, and Galaxy 3 at 93.5° west longitude.

Galaxy 2 and Galaxy 3 will furnish a total of 24 transponders for MCI Communications Corporation. The Galaxy satellites have been designed to take full advantage of the concept of private transponder ownership. Each satellite has 24 transponders, all of which are on a non-common carrier basis. All Galaxy satellites transmit in the C-band frequency; each has a 9-year operating life.

Other Hughes Communications subsidiaries are developing a network of earth stations, which will be used to access the Galaxy satellites in seven major cities. These cities represent the major population, commercial, financial, and telephone traffic centers in the United States.

By mid-1985, the Hughes Group will have a total of seven satellites in orbit; three Galaxy and four Leasat satellites. Galaxy will serve the voice, video and data transmission needs of large corporations, long-haul carriers, cable programmers, and broadcasters. Leasat will provide worldwide tactical communications service to the U.S. Navy, and other agencies of the Department of Defense.

The Galaxy satellites are versions of the Hughes HS-376 series of commercial communications satellites. In addition to Hughes, other users of the HS-376 include A.T.& T., Satellite Business Systems, Western Union, Australia, Brazil, Canada, Indonesia, and Mexico.

The HS-376 is a spin-stabilized satellite, with two cylindrical solar panels—the outer of whic telescopes in space to expose the inner panel to sunlight. When this feature is combined with the satellites folding antenna, Galaxy 1 is only 9 feet high stowed atop its launch vehicle, the Delta 3920 Rocket. After launch, the 7 foot wide satellite reaches a height of 22 feet, after its solar panel is extended and the antenna erected.

In April 1983, MCI Communications Corporation agreed to purchase a total of 24 transponders on the Hughes Communications Galaxy Satellite System. MCI will own 12 transponders on each of the Galaxy 2 and Galaxy 3 satellites, providing voice, data and teleconferencing transmission capabilities.

The transponders will increase the voice and data transmission capacity and capabilities for MCI, enabling the company to meet projected service needs by adding as many as 48,000 more telephone or data

Figure 8.8 *Commodity World News Service requires an Epson type printer.*
Courtesy Commodity Communications, Corp.

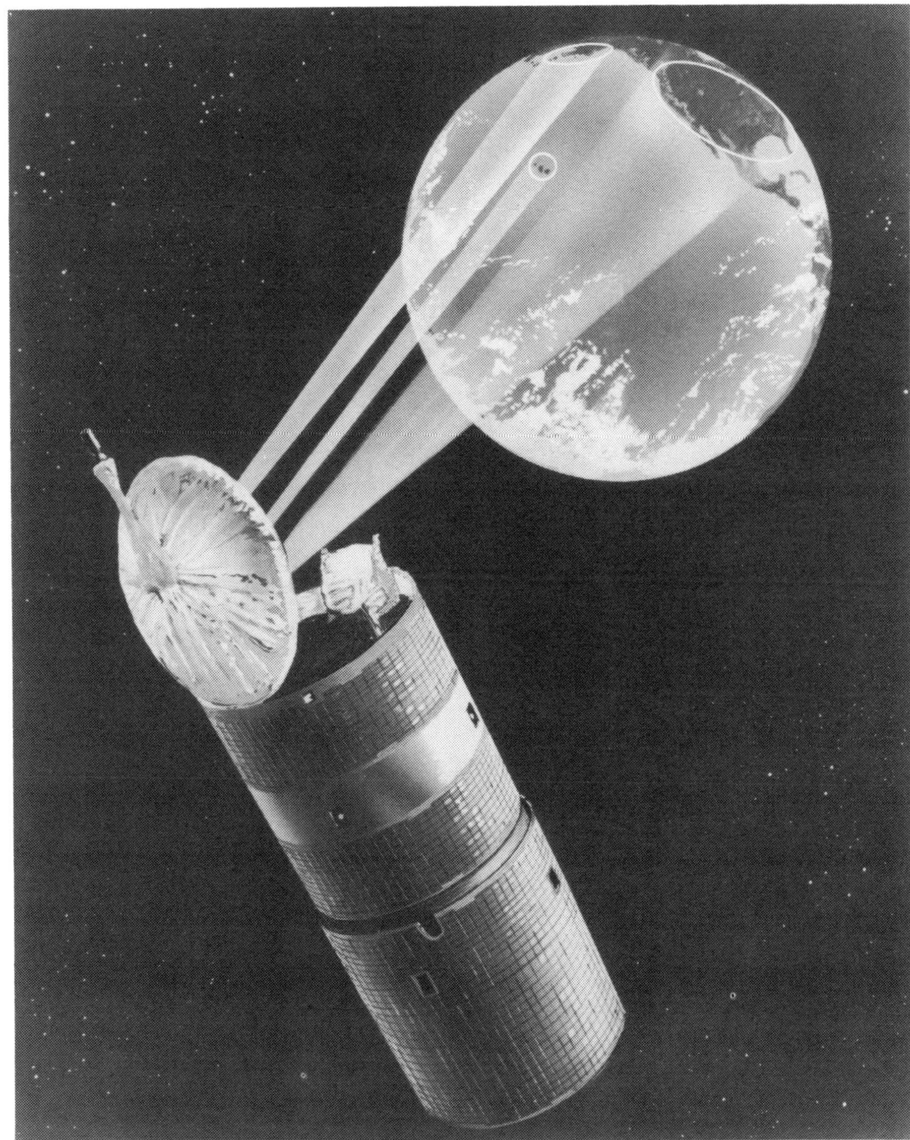

Figure 8.9 *Hughes Galaxy Satellite showing spot beam coverage for data transmission.*
Courtesy Hughes Communications Corp.

Figure 8.10 *UPI's use of the micro dish system.*

circuits. The increase in capacity need will occur when MCI begins receiving access to the rotary telephone market.

MCI will have access to the transponders through Hughes earth station facilities in New York, Los Angeles, Chicago, Dallas, Seattle, San Francisco, and Atlanta, as well as other earth stations to be constructed by MCI.

MCI provides a variety of domestic and international telecommunications services, including long distance telephone services, paging, data services, private line telex and mobile radio.

The Hughes Group has also signed an agreement with Equatorial Communications for the purchase of satellite transponder capacity. This agreement provides for Equatorial's purchase of up to four transponders on Hughes Galaxy 3 satellite. **Fig. 8.9.**

Galaxy 3 transponders will be used with Equatorial's micro earth station product line, which utilizes both 2 foot and 4 foot diameter antennas. The system will use the spread spectrum mode or type of transmission outlined in the previous chapter.

Equatorial's private data networks are currently used for the distribution of general news, commodity, and financial information. The company's customers include: Reuters, Limited; United Press International, Inc., Commodity News Services, Inc., and Market Information, Inc. Over 11,000 of its micro earth stations are in use, and are located throughout the U.S., Alaska, Hawaii and the Caribbean. **Figure 8.10.**

Galaxy 2 and Galaxy 3 are designed to serve the needs of the general business community and other data and information services.

HUGHES SATELLITE SYSTEM DESCRIPTION

System Objectives

The Hughes Communications domestic satellite system provides facilities for C band fixed satellite service to a variety of users within the continental United States, Alaska, and Hawaii. Users include video relay carriers, resale and value added carriers, specialized common carriers, public service groups, private companies, the federal government, and established telephone and telegraph carriers.

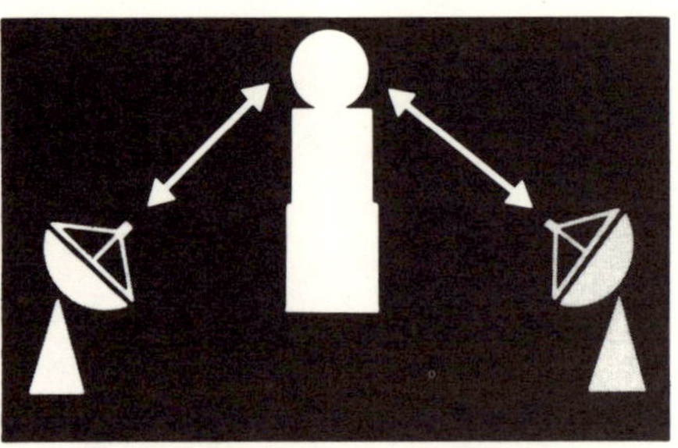

Satellite System

The system consists of two operational satellites at 74° and 135°W longitude, with a third satellite to be launched as user demand requires. All satellites will use the 4 and 6 GHz frequency bands for communications. The initial ground segment consists of two telemetry and command (T&C) receive-transmit earth stations—located at Brooklyn, New York and Fillmore, California—and an operations control center (OCC) at El Segundo, California. Additional receive-transmit and receive-only earth stations will be added to the system as necessary to meet customer demands. **Figure 8.11.**

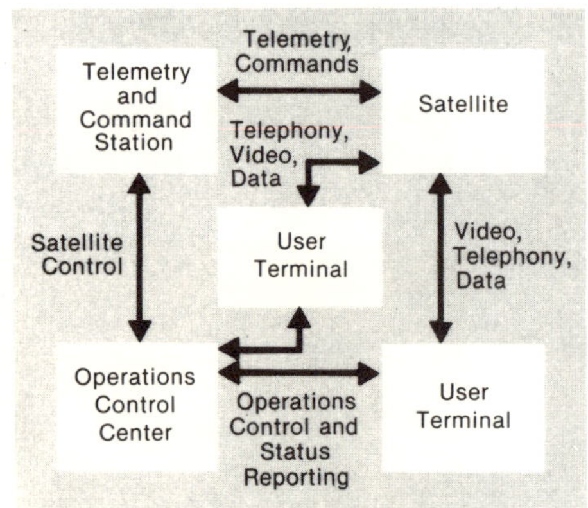

System Operation

Uplink video, telephony, and data signals are transmitted from the user terminals at 6 GHz to the satellite, where they are downconverted and retransmitted at 4 GHz. The OCC controls the satellite through the New York T&C earth station, which transmits commands to the satellite and receives satellite status information. Satellite and transponder telemetry data are analyzed at the OCC, which provides system and transponder status to the users.

Communications Satellite

The satellite is part of the HS 376 communications satellite family developed by Hughes Aircraft Company. It is spin stabilized with a despun payload, and its antenna directs radiated RF power toward the continental United States, Alaska, and Hawaii. Power is generated through a series of solar cells on the exterior of the satellite; power capability is enhanced by the deployment of an extendible solar panel.

Seven HS 376 satellites are now operating. Three of the satellites, launched in November 1980, September 1981, and November 1982, are providing communications services for Satellite Business Systems (SBS). Telesat Canada receives communications from Anik D, launched in August 1982, and Anik C, launched in November 1982. Western Union launched Westar IV in February 1982 and Westar V in June of that year.

Eighteen satellites, including three for Hughes Communications, are under construction. Additional launches are planned for SBS, Telesat Canada, and Western Union.

Three ATT Telstar satellites, which will replace existing Hughes satellites. In addition, the Australian government has ordered three HS 376 satellites for its first national communications program, to begin in 1985. The governments of Brazil and Mexico have also ordered HS 376 satellites for their countries' communications satellite systems.

The HS376 family allows considerable flexibility in communications payload frequency and transponder configuration.

Satellite Performance and Characteristics

The communications payload includes twenty-four 36 MHz wide channels, 12 vertically polarized and 12

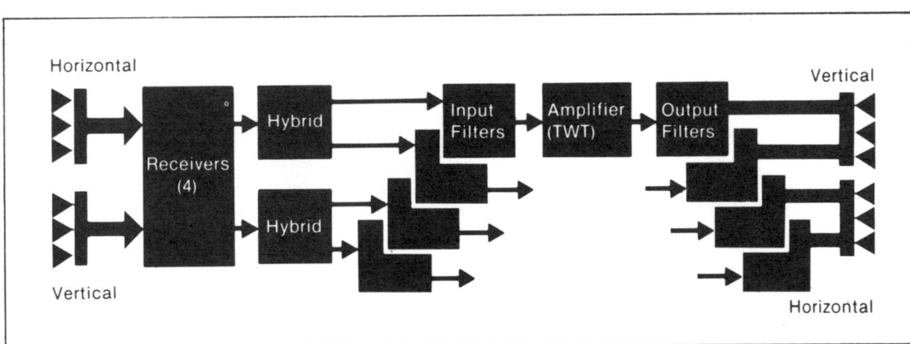

The communications payload uses a horizontally and vertically polarized antenna with associated feeds, receivers, 3 dB hybrid power dividers, input and output filters, and TWTAs with 9.0 watt output power.

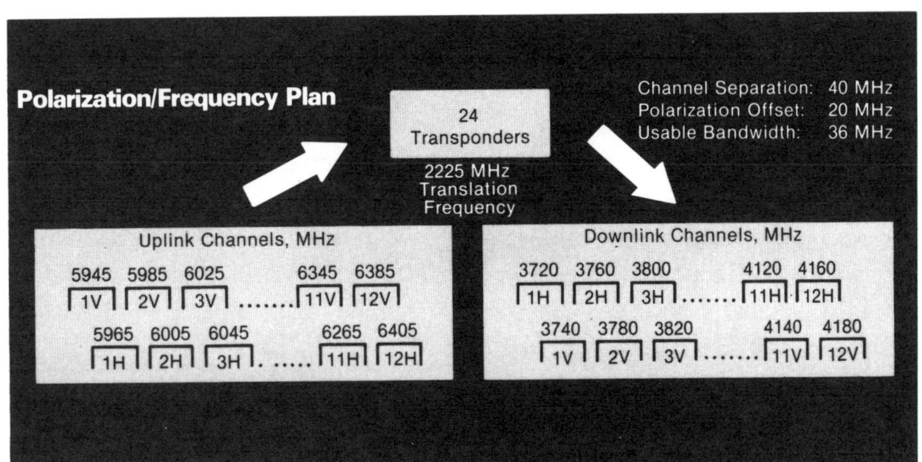

The uplink channel center frequency band is from 5945 to 6405 MHz; the downlink center frequency range is from 3720 to 4180 MHz. Frequency separation between channels is 40 MHz. The offset between vertical and horizontal polarizations is 20 MHz, thereby minimizing interference levels. All uplink channels are translated by 2225 MHz to obtain downlink channel frequencies. Total usable bandwidth per channel is 36 MHz, standard for most U.S. domestic satellite systems.

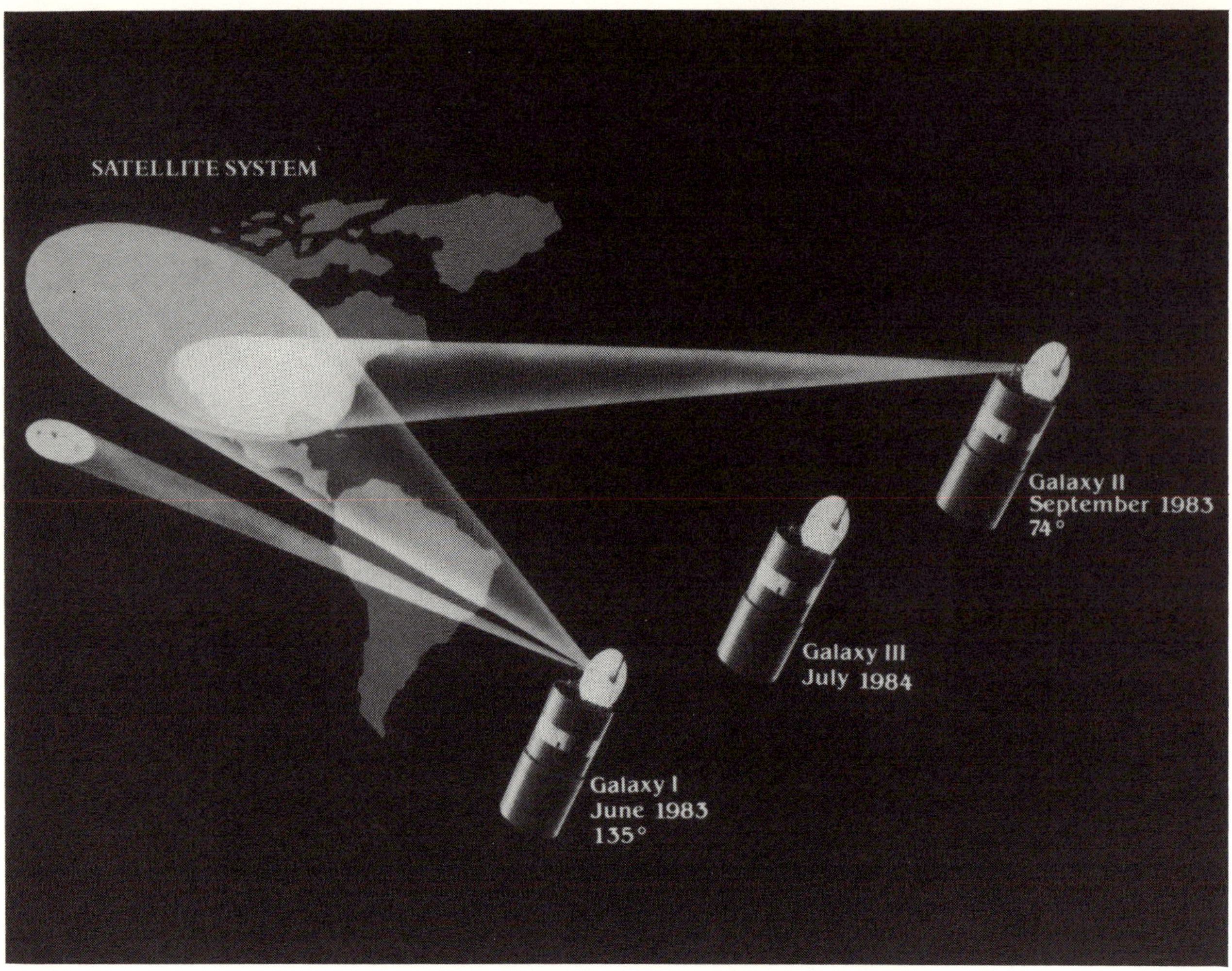

Figure 8.11 *The Hughes Galaxy Satellite System.*
Courtesy Hughes Communications.

horizontally polarized. Each transponder produces 9.0 watts of output power. Six spare traveling wave tube amplifiers (TWTAs) provide the redundancy to increase transponder reliability. The communications antenna is a single reflector with dual polarized apertures. The satellite's mission life, determined by the amount of stationkeeping propellant stored onboard, is at least 9 years; it has a 10 year design life. Through the use of a hydrazine propulsion system, the satellite is maintained on station to within ±0.1 degree in both east-west and north-south directions. A solar cell/battery power subsystem provides sufficient power for full operation during sun eclipse, with 741 watts of dc power available at the end of the satellite life. Satellite attitude is maintained through spin stabilization and a despun platform with pointing accuracy of ±0.05 degree. Satellite mass at beginning of life is 637 kg. An internal solid propellant motor, fired at transfer orbit apogee to place the spacecraft in circular orbit, maintains the satellite at synchronous orbit altitude.

Typical Link Quality

A typical video link from the satellite into a 5 meter diameter antenna will have a peak-to-peak signal-to-weighted rms noise ratio of 53 dB. If a 10 meter diameter antenna is used, signal quality is increased to 58 dB. For digitized video signals, the bit error rate is 1×10^{-6}. Four digitized video channels can be accommodated through one spacecraft transponder (6.3 megabits per second each). For single channel per carrier voice, the test tone-to-rms noise ratio is 35 dB into a 5 meter diameter antenna. With a 10 meter antenna, signal quality is increased to 40 dB.

Service	Quality Definition	Quality	
		10 M Antenna	5 M Antenna
Video	Peak-to-peak signal-to-weighted RMS noise	58 dB	53 dB
6.3 MBPS Digital video	Bit error rate	10^{-6}	—
SCPC voice 1000 channels	Uncompanded test tone -to-weighted RMS noise	40 dB	35 dB

Channels Available at Different Probability Levels

Reliability analyses for the communications transponder payload show a 50 percent probability that 24 transponders will be operational at the end of the 9 year satellite mission life. There is a 90 percent probability that 22 or more transponders will be operational at the end of mission life.

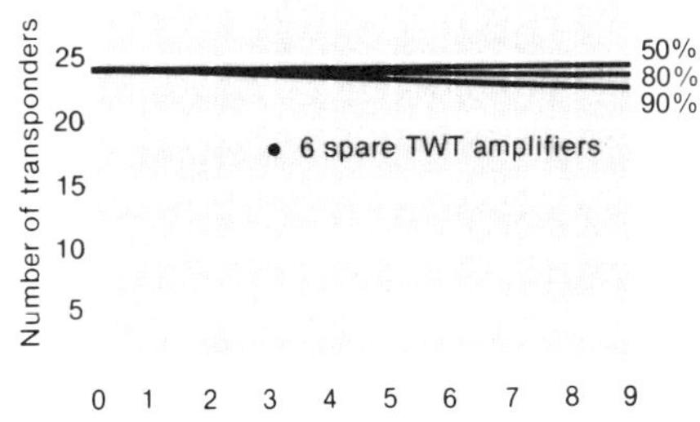

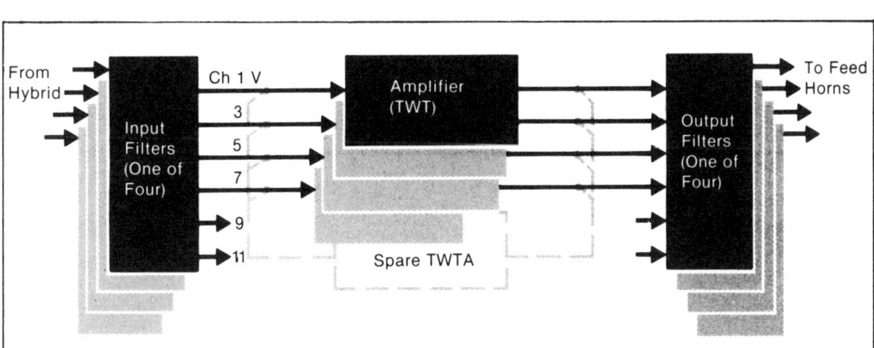

Transponder Redundancy

There are six spare amplifiers in reserve for 24 operating TWTAs. These 24 TWTAs are separated into six groups of four each. Each of these groups is protected by a spare TWTA, which can be used to replace any of the four amplifiers.

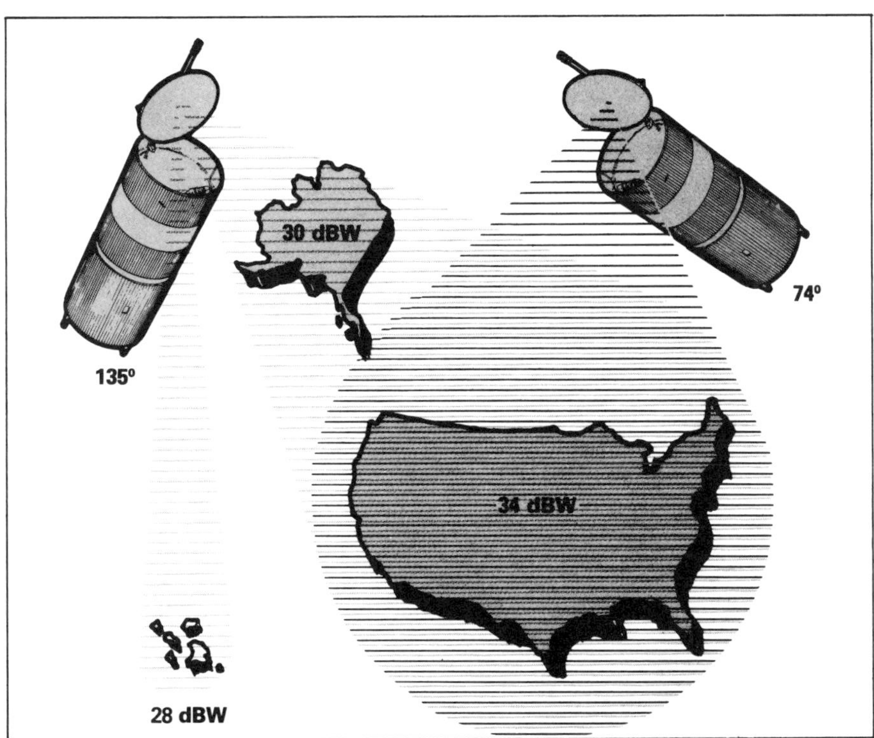

Satellite Communications Performance

The satellite located at 135°W longitude provides 24 operational channels of at least 34 dBW, and 28 dbW to most of the continental United States. The satellite located at 74° provides 24 channels at a minimum of 34 dBW to most of the continental United States; Alaska and Hawaii are not visible from this easterly location.

THE FUTURE

Hughes Communications has applied to the FCC to construct, launch and operate a fourth C-band satellite. Galaxy 4 will feature a technologically advanced transponder design that includes the use of state-of-the-art, high-efficiency solid-state power amplifiers on all 24 transponders in the communications payload.

Looking beyond its currently operating C-band system, Hughes Communications Galaxy has also submitted applications to the Federal Communications Commission to build, launch and operate three new generation satellite systems. The first of these systems will operate in the Ku frequency band, and will feature wide-body spacecraft design. The second system, to operate in the virtually unused Ka frequency band; will provide enormous communications capacity via 16 high power spot beams directed to specific geographical areas.

The third system will operate in the broadcast satellite service band, and will transmit 32 channels of quality video programming directly to individual homes throughout the United States. The development of these advanced systems are designed to complement the existing Galaxy networks.

PRESS SERVICES VIA SATELLITE

(SCPC/FM/FDM)

When tuning over the regular SCPC channels, one will pass over a signal that sounds much like a buzz saw, very raspy and a little wide, and is one of some 19 found on this transponder. Of these, several are radioteletype (RTTY) channels or sections each with approximately 25 channels. Each circuit contains the news service that is directed to a state or a specific geographical area of the U.S. The areas covered are divided again in the heavily populated states and areas. These are known as the Ohio Wire, the Western Pennsylvania-Eastern Pennsylvania Wire, etc. Between transmissions of state or area interest, these channels carry some national, international and sports news.

Reception of this service requires the following equipment:

(1) A small TVRO antenna, 6 foot to 8 foot in size.
(2) A TVRO with a very stable down converter (Drake ESR-240 works well).
(3) A stable FM receiver that is converted to tune from 50 MHz to 85 MHz.
(4) A stable very low frequency (VLF) single sideband (SSB) receiver that will tune from 7 kHz to 18 kHz on upper and lower sideband. (JRC-NRD-515 Receiver is good).
(5) A RTTY demodulator multiplex unit such as the Fredricks 1202RB (a surplus unit).
(6) A video character generator and conversion unit, such as the M-600 unit to furnish video out and printer output.
(7) And, of course, a video monitor for video output, and a printer such as the Gemini 10X Printer, if you want hard copy. This equipment is shown in further detail in **Figure 8.12.**

These narrow bandwidth signals can be received using smaller antennas than found in the average TVRO system. Attention should be paid to the stable down converter used; the output of this down converter should be 70 MHz and not have any hum present caused by the Voltage Controlled Oscillators (VCO) control voltage being placed or run down the coax line from the receivers power supply to the down converter. The proper down converter will carry a separate set of wires going to the down converter from the receiver's rear terminal—any included hum in the audio will effect the final audio quality to the RTTY unit, and will not produce satisfactory results (errors in printout). **Figure 8.13.**

For the FM section, several quality FM receivers that are setup to tune the television audio section (Channels 2 through 5), 55 MHz to 80 MHz can be used. Again, this FM receiver or tuner *must be stable,* and have a bandwidth of approximately 60 kHz. Radio Shack and others sell these receivers and tuners.

After the FM demodulation of this signal to baseband audio, you will have a composite signal consisting of the UPI radio (voice) section, plus several 25 channel FDM radioteletype (RTTY) signals. These will be located at approximately 8.25 kHz and 12.25 kHz. See Chapter 4, last section, Frequency Locations of SCPC Services for FM Frequency.

The subcarrier upper sideband frequencies will be as above (8.25 kHz and 12.25 kHz)—don't get the two frequencies mixed up in the tuning procedure.

The proper tuning procedures will yield you first the voice audio; at this point you will know you are

UPI SATELLITE TO RADIO STATIONS (RTTY) 170 Hz SHIFT

Group One

CHANNEL	SERVICE	BAUD RATE
1	Maine	45
2	Vermont & New Hampshire	45
3	Connecticut	45
4	Rhode Island	45
5	Massachusetts	45
6	New York	45
7	New Jersey	45
8	E. Pennsylvania	45
9	W. Pennsylvania	45
10	W. Virginia	45
11	Ohio	45
12	Kentucky	45
13	Maryland	45
14	Virginia	45
15	North Carolina	75
16	Georgia	45
17	Alabama	45
18	Florida	45
19	Mississippi	45
20	South Carolina	45
21	Tennessee	75
22	Michigan	75
23	Indiana	45
24	Illinois	75
25	Minnesota	45

45 Baud Rate = 60 Words Per Minute

Group Two

CHANNEL	SERVICE	BAUD RATE
1	Nebraska	45
2	N. & S. Dakota	45
3	Wisconsin	45
4	Iowa	75
5	St. Louis	75
6	Missouri	45
7	Oklahoma	45
8	Kansas	45
9	Arkansas	75
10	Louisiana	45
11	Texas	45
12	New York	45
13	Colorado	45
14	Wyoming	45
15	N. California	45
16	California Valley	45
17	Los Angeles	45
18	Arizona	45
19	Nevada	45
20	Montana	45
21	Idaho	45
22	Utah	45
23	Oregon	45
24	W. Washington State	45
25	E. Wash. & N.W. Idaho	45

75 Baud Rate = 100 Words Per Minute

close to the FDM RTTY signals. Careful tuning should bring in the buzz saw sound of the FDM section or channels you are looking for; upper or lower sideband will be used on the general coverage receiver set at either 8.25 kHz or 12.25 kHz. **Figure 8.14.**

If you cannot copy the FDM signal, you should go through a check-out procedure to try and determine where the problem is located. You should check out the following areas:

(1) Make sure you are on proper satellite, Westar 3.

(2) Check for proper transponder (number 1), then check for correct polarity on this transponder—should be horizontal.

(3) Look over the hookup of equipment for correct connections.

(4) Re-check the FM frequencies on the FM section to locate the audio news section—these will appear on the hour and half hour.

(5) Re-check your frequency reading on the general coverage receiver, and use both upper and lower sideband as a further check.

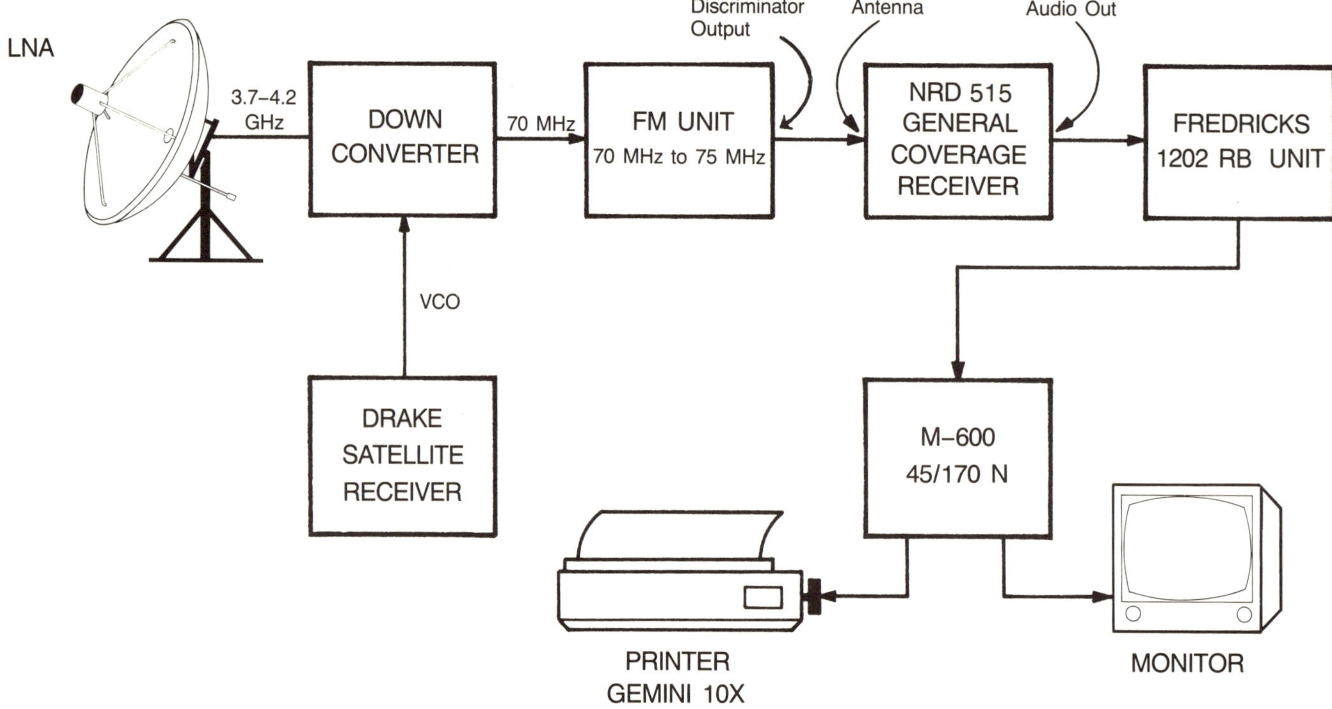

Figure 8.12 *SCPC/FM/FDM equipment diagram using the Fredricks 1202 RB.*

Figure 8.13 *Video read-out of a news service using FDM transmission.*

```
      UPI 05-17-84 11:19 APD
178YR

      URGENT
NERVEGAS

   (WASHINGTON)-- FOR THE THIRD TIME IN AS MANY YEARS, THE HOUSE
HAS

VOTED TO REFUSE THE PENTAGON MONEY TO MAKE NERVE GAS WEAPONS

COMPONENTS. TODAY'S VOTE WAS 247-179. THE HOUSE REFUSED TO ACCEPT
THE

REAGAN ADMINISTRATION REQUEST FOR 95-MILLION DOLLARS TO BUILD WEAPON

PARTS, WHICH WOULD NOT IMMEDIATELY BE ASSEMBLEDIINTO WEAPONS. THE

ADMINISTRATION TRIED WITHOUT SUCCESS TO WIN BACKING FOR THE PLAN
BY

PROMISING THAT THE PENTAGON WOULD ASK CONGRESSIONAL PERMISSION BEFORE

ACTUALLY ASSEMBLING THE PIECES INTO BOMBS AND ARTILLERY SHELLS.

      UPI 05-17-84 11:20 APD
179YR


     WX-05-14-84 1702EDT

      STOCKS-CLOSE-MORE
    -0-
    THE 10 MOST ACTIVE STOCKS IN N-Y-S-E COMPOSITE TRADING...
    STOCK          SALES        LAST  NET CHANGE
    ENSERCH (DIV) 1,344,100  20 3-8 OFF 5-8
    PUB SVC IND      965,800   8 7-8 UP 1-2
    A-T-AND-T CO     959,500  15 5-8 OFF 1-4
    FORD             901,900  34 7-8 OFF 1-4
    DIAMOND SHAM  H 884,800  21 3-8 OFF 1-8
    MOBIL            863,100  28 3-4 OFF 5-8
    SUPERIOR OIL     848,000  41 1-2 OFF 1-4
    EXXON            828,100  42 1-4TLFN1-:
    SOUTHERV        739,000  14 3-4 OFF 3-8
    PHILA ELEC       702,900  13 3-8 OFF 1-4
      UPI 05-14-84 02:05 PPD
241YR

      STOCKS-CLOSE-MORE
    -0-
    THE FIVE MOST ACTIVE STOCKS IN AMERICAN EXCHANGE COMPOSITE
TRADING...
    STOCK       SALES     LAST    NET CHANGE
    COMMuDORE     630,300    1 5-8 OFF 3-8
    TIE COMM      488,500  14 1-8 UP 3-8
    NATL AT DEVL 448,800  17 1-4 OFF 4 5-8
    GALAXY OIL    287,900    3      UP 3-8
    INSTRUM SYST 284,900    2 1-4 UNCH

      UPI 05-14-84 02:14 PPD
31
243YR
```

Figure 8.14 *Print out of World News Service.*

(6) Re-check the proper speed and shift on the Fredericks 1202RB. This should be 45 baud rate (60 words per minute) on 8.25 kHz channels 1–3–6–7–8–9–10–12–13–14–15–16–17–18–19–20–21. Channels 2–4–5–11 will be 75 baud rate, which is 100 words per minute. On the 12.25 kHz channels, the speeds will be 45 baud (60 words per minute) for channels 1 through 4, and channel 18, with channels 15 and 16 at 75 baud or 100 words per minute.

(7) The correct shift and phasing of the signal should be checked on the Frederick's unit—170 Hz—normal phasing.

(8) Complete overall check of system.

Full information on some of this equipment can be obtained by writing the author, Thomas P. Harrington, at the publisher's address listed in the front section of this book. Please enclose a stamped, self-addressed #10 envelope.

"COOP'S SATELLITE DIGEST" features a monthly section on SCPC data and audio services. A subscription is $75 per year for 12 monthly copies, and 12 CSD/2 copies, which are published in the middle of the month. Both issues are mailed *first class*. Contact "COOP'S SATELLITE DIGEST," P.O. Box 100858, Ft. Lauderdale, Florida 33310. Phone: 305–771–0505.

Please return the up-date coupon at the rear of the book to have your name added to our up-date list for future information. Enclose a stamped, self-addressed #10 envelope.

In **Figure 8.15,** we show a data system using a new multiplex unit, along with the FM section, and an M-600 multi-code unit to receive the data stream.

OTHER FDM/RTTY LOCATIONS

Westar 2, transponder 7—various services, 110 baud ASCII

Westar 2, transponder 9—various services Baudot, 57 baud

Galaxy 2, transponder 18—various services, all speeds

Telstar 301, transponder 16—various RTTY channels

Satcom 2R, many—various data systems

Plus satellites shown in the list, Chapter 4, Where to Find the SCPC Services. Any transponder listed as data, or SCPC, can contain FM/FDM/SCPC signals. Other information will be published in CSD and CSD/2, also from up-date, information from the publisher. (See back page for mail-in coupon.)

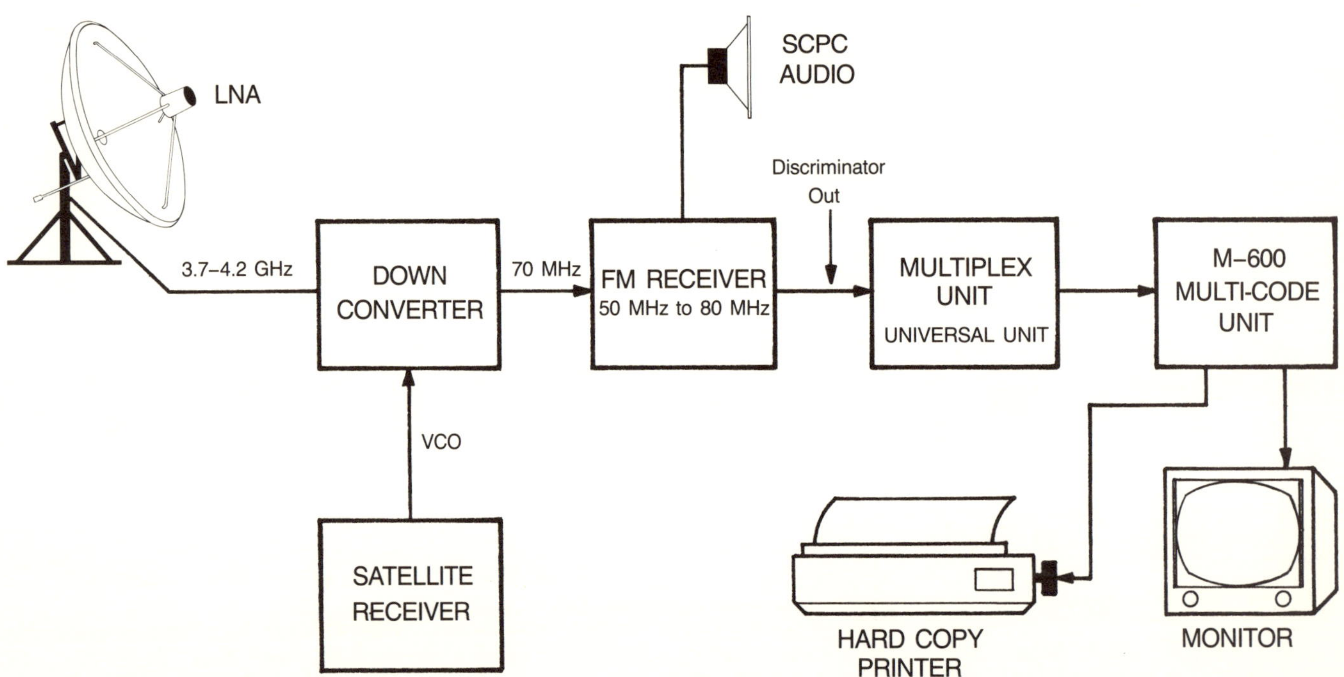

Figure 8.15 *Equipment for FM/FDM data reception with Universal Multiplex Unit and M-600 Multi-code Converter Unit.*

EARTH STATION GLOSSARY OF TERMS

ADJACENT CHANNEL INTERFERENCE: Signal distortion caused by signals in nearby frequencies which are not properly filtered. Use of a special bandpass filter allows only the selected frequency band to pass through it removing adjacent channels.

AFC AUTOMATIC FREQUENCY CONTROL: A satellite TV receiver feedback circuit which prevents the tuning oscillators from drifting away from the center frequency of the selected channel due to temperature change or other instabilities. A phase lock loop demodulator usually provides an AFC output back to the Local Oscillator.

AGC AUTOMATIC GAIN CONTROL: A satellite TV receiver feedback circuit which controls the gain (amplification) of the IF amplifiers so that the signal input to the demodulator will be constant. despite incoming signals of varying strength from different satellite transponders. AGC can be overridden by a manual gain control to make signal strength measurements in most receivers.

ALIGNMENT: The process of tuning (or tweeking) a circuit to compensate for the approximate tolerances of the components during assembly. using test equipment.

AM AMPLITUDE MODULATION: An easy method of transmitting program signals on a carrier frequency where the relative strength (amplitude) of the carrier is made proportionally equal to the amplitude of the program signal. AM is simpler but more susceptible to noise than FM. Satellite TV uses FM for both audio and video modulation but the user's TV set takes the satellite TV receiver's output and detects the video as AM. the audio as FM.

ANIK: The Canadian domestic satellite system used to transmit the Canadian Broadcasting Corp.'s network TV feeds. All ANIK satellites are operated by TeleSat Canada of Ottawa. ANIK satellites have both 4GHz C-band and 12GHz Ku-band transponders.

APERTURE EFFICIENCY: The ratio of captured signal to the theoretical maximum for a given dish antenna-feed combination. The design goal is 100% aperture efficiency, but most TVRO dishes perform at only 50-80% to attain low noise characteristics and ease of construction. Some VHF/UHF antennas on the other hand can approach the 100% goal with an array of reflective elements.

ARO AUDIO RECEIVE ONLY: Small dish antennas used by radio networks for music and news programming distribution from TV satellites (mostly WESTARS). Dishes 2 meters and smaller have been considered by radio broadcast stations.

ATTENUATOR: A passive device which causes a known insertion loss in the signal transmission line. It is commonly used to prevent a very strong signal from overloading a receiver.

AUDIO SUBCARRIERS: The sound in a TV satellite composite signal is encoded in a narrow portion of the video carrier, usually a high fidelity FM signal at 6.2 or 6.8MHz as measured after the main signal has been demodulated. Other satellite subcarriers can carry digital and test information as well.

AZIMUTH: Angle from the north to south line through the earth station's location. Measured in degrees clockwise.

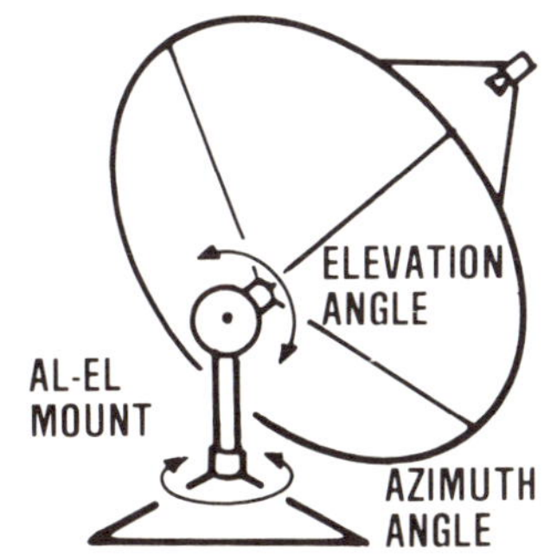

AZIMUTH-ELEVATION MOUNT: A movable dish with an AZ-EL mount allows rotation in the horizontal-plane from the azimuth angle from the due north with one adjustment. Second adjustment provides elevation adjustment.

BANDPASS FILTER: A type of electronic frequency filter which severely limits signal frequencies above and below the selected frequencies, preventing adjacent channel interference. Satellite TV receivers use these also to remove noise from around the edges of the selected channel, usually 30MHz wide.

BASEBAND: This is the output signal of a video camera, videotape recorder or satellite TV receiver before remodulation (so that it can be viewed on an ordinary TV set). A signal in a satellite TV receiver goes from 4GHz through the downconverter to become IF and then through an FM demodulator to become baseband. American NTSC TV band width is 4.2MHz at baseband.

BEAMWIDTH: The beamwidth of a dish antenna is the angle of sky which can be illuminated (picked up or sent out) by the dish. Within that arc satellites can be seen from the TVRO dish. Large dishes have narrow beamwidths which reduce noise from its sides. Small dishes have wider beamwidths and are noisier, but easier to aim.

BIPOLAR: A type of silicon transistor used in LNAs and other high-frequency low-noise devices. They are superior in noise quality to ordinary transistors but are inferior to FETs especially GaAsFETs.

BNC CONNECTOR: Easy to lock coaxial cable fittings which interface signals in the IF portions of a satellite TV receiver. They work well in the 70MHz range.

BORE SIGHT: The center of the transponder "Footprint" where a signal strength is at its maximum.

C BAND: Designation for 3.7 - 4.2GHz microwave frequency band used for the downlinks of satellite TV signals. Wavelengths are between 8.10 and 7.14 centimeters (3.19 and 2.81 inches).

C/N: Carrier-to-Noise ratio. The ratio of the carrier strength and noise strength measured in dB. The higher the C/N, the higher the S/N and quality of the TV picture. 11dB is considered excellent. Above 7dB is good.

CARRIER: A radio wave of constant amplitude, frequency, and phase at the frequency of operation for the satellite communications system. Uplink carrier is 5,900 to 6,400MHz. Downlink carrier is 3,700 to 4,200MHz.

CASSEGRAIN FEED: A radiating system which includes the primary reflector (earth station dish) and a secondary reflector and feed horn.

CHANNEL: A frequency band allocation which defines the limits of the contained broadcast carrier signal. In the USA channels are allocated by the FCC.

CLOSE CAPTIONED TV: A text service for the hard-of-hearing TV audience which decodes a text subcarrier and displays it at the bottom of the TV frame on the accompanying video picture. It does not interfere with the standard audio FM subcarrier.

COMMON CARRIER: An operator or leasor of satellite TV transponders which in turn leases them to other parties, or transmits programming for others without controlling or owning the content. 4GHz satellite TV is not legally a broadcast service and the FCC does not make the satellite TV common carriers (RCA Americom, Western Union. . .) abide by the constraints of broadcasting law.

COMPARATOR: In an FM demodulator using a phase lock loop (PLL), this is the electronic component which compares the phase relationship of the input signal with the signal from the tracking local oscillator (LO). The output signal from the comparator is proportional to the phase error between the two input signals and is used to control the LO.

COMPOSITE TV SIGNAL: Combination of video picture, color, audio and synchronization information.

dB AND dBW: dB is the standard unit for expressing relative power, voltage, or current, dBW indicates actual power of one signal compared to a reference of 1 watt.

DBS: Direct Broadcast Satellite, a technology used to transmit electronic signals directly from a communications satellite to the end user. The Federal Communications Commission has set aside frequency space in the 12/14GHz Ku-band for DBS services in the U.S.

COMSTAR: A series of C-band domestic satellites owned by the Communications Satellite Corp., a company created by an act of Congress in 1962 to operate the international satellites for the INTELSAT consortium of countries. Comstar satellites are currently leased to AT&T for television and other transmissions.

DETECTOR: A demodulator circuit in a satellite receiver which extracts the program signal from the carrier.

DIPOLE: An active antenna element located in the feed which collects the satellite TV signal and directs it to the LNA.

DISCRETE COMPONENTS: Assembly technique in which each part is built separately and then assembled.

DISH ANTENNA: A dish-shaped metallic or metallic-covered device which is used with a transmitter or receiver to amplify electronic signals relayed into or collected from the atmosphere.

DOMSATS: Domestic communication satellites including SATCOM, COMSTAR, GALAXY and WESTAR.

DOUBLE CONVERSION: This downconversion technique converts from 4GHz to the final IF (typically 70MHz) in two stages instead of just one, so that potential image noise from the first mixer stage is eliminated.

DOWNLINK: (Satellite to earth) Satellite programming is relayed via a downlink transmitter on the satellite back down to earth and spreads to cover a vast area. The satelite to earth transmission is carried at 3,700 to 4,200MHz.

DOWN CONVERTER: Device used to convert the 3.7-4.2GHz microwave downlink signal down to a VHF frequency range typically 70MHz.

DYNAMIC RANGE: The weakest through strongest signals that a receiver will accept as input. Signals which are too weak cause excess noise and signals which are to strong cause overloading and possibly modulation distortion.

EIRP: Effective Isotropic Radiated Power. A measure of the relative strength of the satellite TV signal expressed in dBW. EIRP figures range from 30dBW to 37dBW.

EARTH STATION: The complete satellite TV receiving system. Also referred to as TVRO (television receive only) system.

ELEVATION: Angle above the horizon measured in degrees. Zero is the horizon and 90° is directly overhead.

FEEDHORN: A component on a satellite TV receiving antenna which collects the signal reflected from the dish surface and "feeds" it into the low noise amplifier.

FET FIELD EFFECT TRANSISTOR: A low-noise high-frequency transistor amplifier which has a current source, gate and drain. The gate is a voltage controlled resistor which regulates the power flowing from the source to the dish.

FIXED SATELLITE: A form of international frequency band allocation where all the sending (uplink) and receiving (downlink) stations are identified. This is the current status of the 4GHz TV satellite systems. See also Broadcast Satellite.

FM IMPROVEMENT: The potential noise reduction in an FM signal due to the demodulation process in a satellite TV receiver. This figure is at most 38.6dB, and is attained above the FM threshold. Below this point it rapidly drops from 37.6dB. Above threshold: S/N = CN + 38.6dB.

FM THRESHOLD: An input signal level which is just enough to enable the demodulator circuits to extract a good picture from the carrier. With test equipment, static threshold is the point at which S/N drops more than 1dB from the straight graph line: S/N = C/N + 38dB. Typically FM threshold is 7-8dB in a satellite TV receiver with threshold extension.

FOOTPRINT: A signal strength map showing the EIRP contours of equal signal from a TV satellite transponder on a given part of the earth's surface.

FREQUENCY AGILE: This is a feature of satellite TV receivers which enables them to tune in all the 12 or 24 channels from a satellite. Receivers sold without this feature are dedicated to one channel and can be tuned by switchable crystals.

FREQUENCY COORDINATION: A service which implements a computer and a USA database to provide the user with information concerning whether there are undesired microwave signals that can affect the operation of satellite television receiver, to communicate with other microwave users to see if they have any planned stations which might later interfere with and disrupt operations of the earth station, or assist in FCC licensing, to register the earth station particulars so that future microwave users can avoid interference.

FREQUENCY MODULATION (FM): A method of transmitting program material which is more interference-free than AM. The frequency of the carrier signal is made proportional to the amplitude of the program signal.

FRONT-TO-BACK RATIO: The ratio in dB of the antenna gain in the forwards direction to the antenna gain in the rear direction. It is a measure of the noise potential from the rear.

G/T GAIN OVER NOISE TEMPERATURE. A TVRO measure of quality expressed in dB. The higher this figure, the better the system. It can be improved by increasing gain or by decreasing the system noise. G/T (degrees Kelvin) = Antenna gain/log (antenna noise temperature + LNA noise temperature).

GaAsFET GALLIUM ARSENIDE FIELD EFFECT TRANSISTOR: This low noise device, although expensive, is used in the highest quality LNAs. The term is pronounced gasfet.

GALAXY: A high-powered C-band satellite system owned by Hughes Communications.

GEO-SYNCHRONOUS ORBIT: The path that the DOMSATS are in at 22,300 miles above earth's equator. Travelling at 7,000/mph their speed and orbit are matched exactly to the earth's rotation and thus are stationary relative to the earth. This allows a fixed TVRO installation.

GHOST: One or more dim copies in a TV picture caused by reflected VHF or UHF broadcasts. Also called multipath distortion, this is not present in satellite TV signals because extremely directional dish antennas are used.

GHz GIGAHERTZ: The standard abbreviation for billions of cycles per second. 3.7 - 4.2GHz is the microwave frequency band allocated for satellite TV in the USA.

GUARD CHANNEL: Unused portions of the frequency spectrum which are located between program channels to prevent adjacent channel interference.

HARMONICS: Spurious signals produced by an oscillator circuit which occur at integral multiples above the resonant frequency of the oscillator. They appear like overtones on a single piano note. They can cause design problems in a receiver circuit unless proper filters are used to remove unwanted harmonics.

HEADEND: The point on a signal distribution system where UHF/VHF/FM and satellite TV signals are captured, combined and fed into the system.

HORN ANTENNA: A type of satellite TV antenna which is shielded against sidelobe interference. The incoming signal is reflected 90 degrees into a cone shaped feed horn. They are much more expensive than a dish antenna of the same aperture.

HPA: High Power Amplifier. A device which provides the energy for carrier amplification necessary to transmit to the satellite.

IC INTEGRATED CIRCUIT: A a solid state complex device which is mass produced on single silicon chips.

IF INTERMEDIATE FREQUENCY: For satellite TV receivers this is usually 70MHz and is the frequency at which most of the signal processing takes place because the design is simplified, and 70MHz parts are less expensive than 4GHz equivalents.

IF STRIP: A PC module which amplifies and filters the output signal of the downconverter in a receiver and inputs it to the FM demodulator. Its gain is controlled by the AGC circuit.

IMAGE NOISE: When a signal is downconverted using a mixer and LO, noise can be passed through the system that is on the mirror image frequency from the selected channel with the LO frequency as the point of symmetry, subsequent bandpass filters remove this noise in double conversion downconverters. A preselecter filter in single conversion receivers does the same thing.

IMPEDANCE MATCHING: The design of a signal interface such that the signal transmitted through it is maximized and the reflected signal is minimized. Standard impedance for LNAs is 50 ohms and for satellite IF circuits 75 ohms. Most signal distribution systems interface at 75 ohms impedance.

ISOLATOR: A device which is a one way valve for microwave signals which prevents stray receiver signals from leaking out past the LNA onto the antenna. It also facilitates the design of the LNA by impedance matching the feed probe to the first LNA amplifier stage. Most LNAs have an isolator attached between the CPR-229 feed flange and the main amplifier box.

KELVIN: The absolute Celsius temperature scale 0° Kelvin is equal to -273.1°C, or absolute zero. An LNA is rated by the amount of noise it contributes to the television picture. A perfect LNA with no noise would have a noise factor of 0 degrees Kelvin. Typical TVRO LNA's range from 120°k to 85°k.

LATITUDE: TVRO site latitude measurements given in degrees, minutes and seconds. Latitude numbers indicate how far north or south a point is.

LNA: Low Noise Amplifier that boosts the satellite signal picked up by the earth station dish. It is installed at a particular spot on the dish and is directly connected to the feed-horn.

LNB: Low Noise Block downconverter, a TVRO component combining an LNA with a block downconverter. (A block downconverter simultaneously converts the entire frequencies of one polarity received by the TVRO to a lower frequency.)

LNC LOW-NOISE CONVERTER: A 3.7 to 4.2GHz low noise converter is designed to mount at the prime focal point of the earth station antenna or at the rear of the antenna. LNC combines amplification and downconversion down to 270-770MHz range. Eliminates need for expensive coaxial cable such as used with LNA configuration.

LOBE: An area of strong reception in a graph of antenna gain versus angle off boresight. In highly directional dish antennas the front lobe is high gain and the side and back lobes are much weaker.

LOOK ANGLE: Angle above the horizon at your location from which the satellite signal arrives.

LONGITUDE: TVRO site longitude given in degrees west or the Greenwich, England longitude line for positive entries and negative entries for east of the line.

LOW PASS FILTER: A circuit which features high impedance of relatively high frequencies and low impedance for low frequencies, in effect blocking the high frequency component in a signal. *See also High Pass Filter and Bandpass Filter.*

MARGIN: A 3dB increase over the FM threshold, in a TVRO system, of the C/N which will provide an output signal completely free of impulse noise (sparklies). The FCC set this figure when CATV earth stations were more strictly regulated.

MHz MEGAHERTZ: The standard abbreviation for millions of cycles per second. Henri Hertz was a famous scientist who pioneered wireless communications.

MIXER: That part of a downconverter which joins together an input signal with that of an LO (local oscillator) to create an output signal which has a frequency that is the numerical difference between the two input frequencies. The program material in the FM carrier is not affected by the mixing process to a lower frequency.

MULTISTAGE LNA: Three or more transistor amplifier stages are placed end to end (cascaded) so that the gain contribution of each one will add up to approximately 50dB. In most LNAs the first stage, closest to the antenna feed probe, has the best noise characteristics to minimize the noise propagated along through the remaining stages.

N CONNECTORS: Coaxial cable fittings which interface cables between the LNA and satellite TV receiver. They carry 4GHz signals at 50 ohm impedance with low loss.

NF NOISE FACTOR: This measures the thermal noise contribution of LNAs and receiver front ends. LNAs typically measure 1.5dB (120 K) and receivers average 11-15dB.

P-P: Peak-to-Peak measurement of a waveform's amplitude as opposed to RMS (Root Mean Square) which is otherwise implied.

ORTHOMODE COUPLER: A section of the waveguide that allows two LNAs to be attached at right angles to each other and simultaneously receive both horizontal linear and vertical linear polarized satellite TV signals such as those that are transmitted by SATCOM.

PARABOLIC DISH: This antenna shape is commonly used to focus all the microwave energy collected on the dish surface into a single point in front of the dish called the focal point. It can be built from metal sheets, fiberglass, mesh or wood, covered with a metal reflecting surface which is accurate, with respect to a perfect parabola, to within 0.1 inch.

PLL PHASE LOCK LOOP: One form of FM demodulator which employs a feed back loop to lock a local oscillator to the same frequency and phase as the input signal. The error corrections applied back to the LO are equal to the original program signal that modulated the carrier and they are passed through the circuit as the output.

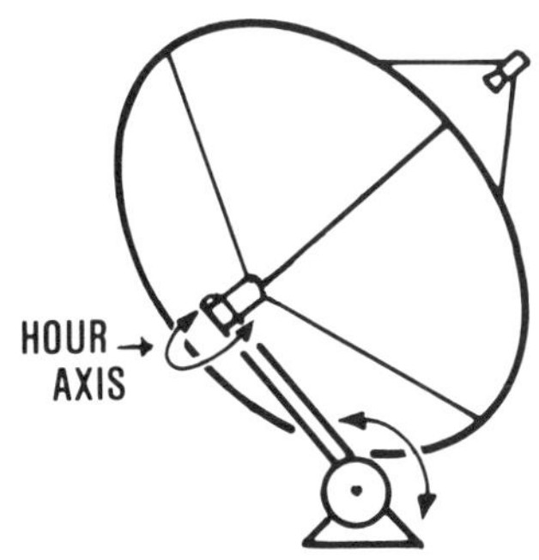

POLAR MOUNT: An earth station antenna mount used to aim reflector while making one adjustment. The hour axis is positioned to sweep the satellite arc. This mount is easier to aim than the azimuth-elevation mount.

POLARIZATION: Satellite signals that are polarized to oscillate only in one plane - horizontal or vertical. Two adjacent channels can be broadcast on overlapping frequencies simultaneously without interfering with each other if the signal for one is polarized vertically and the carrier wave for the other is polarized horizontally. Satellites that offer 24 channels of programming use this principle to provide 24 channels of

PORT: A physical signal interface. This can be a waveguide flange or connector.

POWER DIVIDER: A passive device similar to a signal distribution splitter but it is used at 4GHz to split an LNA's signal for up to 12 separate channels (either all vertical or horizontal). A full channel CATV or MATV system would require two LNAs and two power dividers to drive 24 separate receivers.

PRE-EMPHASIS: A selective amplification of the high frequency end of a satellite TV channel prior to uplink transmission to overcome potential noise problems. Each satellite TV receiver reverses the transformation in its de-emphasis circuits.

PRIME FOCUS FEED: The section of the antenna that receives the reflector and funnels the signal to the LNAs for amplification.

PROBE: The driven element in a microwave dish antenna system. It is located in the feed and converts RF energy in the waveguide to a signal on a transmission line to the LNA.

QUIETING CURVE: A graph of the signal-to-noise ratio (S/N) versus the carrier-to-noise ratio (C/N) for a particular satellite TV receiver. Generally for C/N above 8dB S/N=C/N + 38. The point on the curve below 8dB C/N where S/N rapidly falls off is the FM threshold.

REGISTERED TVRO: The FCC accepts registered earth stations so that it can be protected from possible terrestrial interference. Frequency coordination must be performed as part of the registration procedure.

RF: Radio Frequency. Any frequency between an audio sound and the infrared light portion of the spectrum, usually considered to be 10MHz to 10,000GHz.

RFI RADIO FREQUENCY INTERFERENCE: Any electrical spurious signals in the IF range causing static and noise in a receiver. RFI can also be caused by improperly shielded components within a receiver.

ROTOR SYSTEMS: A method of rotating an LNA feed 90 degrees to switch between vertical and horizontal polarizations. Many times an antenna rotor can be used with remote control from indoors.

S/N: Signal-to-Noise ratio. A ratio of the magnitude of a signal to that of the noise.

SATCOM: A group of U.S. C-band communications satellites owned and operated by RCA Americom.

SATELLITE RECEIVER: A microwave frequency range television receiver. Removes audio and video information off the 70MHz carrier signal. Provides additional amplification and selects transponders. Features transponder (channel select) control.

SCRAMBLING: Techniques to encipher a TV signal to prevent unauthorized reception without a descrambler device. Typically this is done by coding the sync information of the video signal.

SHROUDING: Protective walls or screens around a dish antenna which stop side interference. It is not needed at most earth station sites.

SIDELOBES: Areas from which noise can leak into a dish antenna from the side. Sidelobe performance is the ability of a given dish to reject these in favor of the satellite signal.

SIGNAL COMBINER: This is the reverse of a signal splitter. It allows several TV signals on different channels to be merged onto a single broadband transmission line. Many times this device can be substituted for by a signal splitter connected in the reverse direction.

SIGNAL SPLITTER: This is a passive device which enables two or more TV sets to divide a TV signal between them with proper balancing and isolation. It can be supplied in either 75 or 300 ohm impedances.

SINGLE CONVERSION: This technique uses just a single local oscillator and mixer to convert a satellite TV signal from 3.7 - 4.2GHz down to the final IF (usually 70MHz). Lower parts count and ease of assembly are important advantages over double conversion but care must be taken to prevent noise on the image frequency from leaking into the output.

SKY NOISE: Background microwave radiation coming from deep space which can be a noise source for dish antennas. Sky noise provides a lower boundary for the possible noise temperature of any dish antenna and is approximately 16-20K.

SMATV: Industry term for a Satellite Master Antenna System which distributes satellite signals throughout an apartment, hotel, motel, condos, etc.

SNOW: Dot type TV interference associated with weak signals in UHF/VHF TV pictures. *See also Sparklies.*

SPACE ATTENUATION: The loss in a TV satellite signal due to the fact that the beam spreads out after leaving the antenna. This is a major factor in path loss.

SPARKLIES: Signal noise which appears on the TV screen as white or black flashing dots. Indicates lack of signal or microwave interference.

STABILITY: The ability of a tuning circuit to avoid drift that most often is caused by ambient (surrounding) temperature changes. Lack of stability is the main reason that AFC circuits are used in satellite TV receivers. Crystal control provides the best stability. The term is also used to describe the ability of an amplifier to resist feedback of the output signal around to the input side.

SUBCARRIER: A second signal "piggybacked" onto a main electronic signal. In satellite TV applications, the video is transmitted via the transponder's main carrier while the audio goes out over an FM frequency subcarrier. Some satellite transponders carry up to four audio or data subcarriers, which carry information which can be related or unrelated to the main signal's programming.

TEARING: A form of weak signal interference which causes ragged streaks on the TV picture in vertical lines joining light to dark transitions. If this occurs in a satellite TV picture it is a good indication that the receiver is operating well below FM threshold.

TERMINATION: A connector or passive device at the end of a signal transmission line. This is like an end cap to maintain the impedance of the line.

TERRESTRIAL MICROWAVE: Communications links on the ground using microwaves. One of the allowed ground frequencies is the same as the band allocated to TV satellites and frequency coordination is needed by commercial TVROs to resolve conflicts.

THRESHOLD EXTENSION: A circuit technique sometimes located in the loop filter of a phase lock loop demodulator, which improves the low signal performance of a receiver by lowering the FM threshold by 3dB C/N.

TRANSPONDER: Separate channels whthin the band of frequencies 500MHz wide (3,700 to 4,200MHz and 5,900 to 6,400MHz). One transponder occupies 40MHz of space. On ANIK and WESTAR satellites there is space within this 500MHz band for 12 channels on SATCOM and COMSTAR satellites. The manufacturers designed these birds so two separate channels or transponders share the same band of frequencies thus doubling the total of 24.

TRUE NORTH: Geographic north, determined by the earth's axis, as opposed to magnetic north, which is determined by the earth's magnetic poles. True north is one of the measurements used to align TVRO dish antennas with the North American satellite belt.

TV MODULATOR: Takes video and audio outputs of satellite receiver and converts baseband video and audio into a pattern which the standard TV receiver can use. TV channel position is usually set on Channel 3 or 4.

TVRO: Television receiver only. Industry term used to describe earth stations designed to receive satellite signals.

UPLINK: (Earth to Satellite) Satellite programming originates in a studio and is carried to a large transmitting antenna which is aimed precisely at the location in space where the desired satellite is in the geostationary orbit. The earth to satellite transmission is carried at 5,900 to 6,300MHz.

VHF VERY HIGH FREQUENCY: TV channels 2 through 13 in the following bands; channels 2-4 occupy 54 through 72MHz, channels 5-6 occupy 76 through 88MHz, and channels 7-13 occupy 174 through 216MHz.

VSWR VOLTAGE STANDING WAVE RATIO: A measure of the efficiency of a signal interface, especially the impedance match of the antenna to the LNA.

WAVEGUIDE: A transmission line consisting of a hollow conducting tube within which electromagnetic waves are propagated.

WESTAR: The domestic C-band satellite system owned by Western Union.

COURTESY OF

SATELLITE COMMUNICATIONS DIVISION

SATELLITE DATA C–BAND
PAST, PRESENT AND FUTURE

Position (West Long.)	Satellite	Operator	Launch Date	Number of Transponders
41	TDRS 1	Space Communications Co.	April 1984	(all leased to NASA)
67	Satcom VI	RCA Americom	May 1986	24
69	Spacenet II	GTE Spacenet	September 1984	18
72	Satcom IIR	RCA Americom	September 1983	24
74	Galaxy II	Hughes Communications	September 1983	24
76	Telstar 302	AT&T	August 1984	24
76	Comstar D1/2	Comsat (AT&T leases)	July/May 1976	24
79	Westar II	Western Union	October 1974	12
81	ASC 1	American Satellite Corp.	September 1985	24
83	Satcom IV	RCA Americom	January 1982	24
86	Telstar 303	AT&T	August 1984	24
87	Comstar D3	Comsat (AT&T leases)	June 1978	24
91	Westar III	Western Union	August 1979	12
93.5	Galaxy III	Hughes Communications	May 1984	24
96	Telstar 301	AT&T	July 1983	24
99	Westar IV	Western Union	February 1982	24
104.5	Anik D1	Telesat (Canada)	November 1982	24
109	Anik D2	Telesat (Canada)	November 1984	24
109	Anik B1	Telesat (Canada)	November 1978	12
114	Anik A3	Telesat (Canada)	May 1975	12
119	Satcom II	RCA Americom	March 1976	24
122	Spacenet I	GTE Spacenet	April 1984	18
123	Westar V	Western Union	June 1982	24
127	Comstar D4	Comsat (AT&T leases)	February 1981	24
128	ASC 2	American Satellite Corp.	September 1986	24
131	Satcom IIIR	RCA Americom	November 1981	24
134	Galaxy I	Hughes Communications	June 1983	24
136	Satcom I	RCA Americom	December 1975	24
139	Satcom IR	RCA Americom	April 1983	24
143	Satcom V	RCA Americom	October 1982	24
171	TDRS 2	Space Communications Co.	late 1984	(all leased to NASA)
TBD	Spacenet III	GTE Spacenet	early 1985	24

APPENDIX–A

NORTH AMERICAN C-BAND SATELLITE FREQUENCY/TRANSPONDER CONVERSION TABLE

Channel or Dial Number	Uplink Frequency in MHz	Downlink Frequency in MHz	Satcom 1 & 2 1R, 2R, 3R 4 & 5	Telstar 301 Comstar 1 & 2, 3 & 4	Anik D	Galaxy 1 & 2	Westar 4 & 5	Spacenet 1 Westar 2 & 3 Anik 3 & B	Channel or Dial Number
1	5945	3720	1 (V)	1V (V)	1A (H)	1 (H)	1D (H)	1 (H)	1
2	5965	3740	2 (H)	1H (H)	1B (V)	2 (V)	1X (V)		2
3	5985	3760	3 (V)	2V (V)	2A (H)	3 (H)	2D (H)	2 (H)	3
4	6005	3780	4 (H)	2H (H)	2B (V)	4 (V)	2X (V)		4
5	6025	3800	5 (V)	3V (V)	3A (H)	5 (H)	3D (H)	3 (H)	5
6	6045	3820	6 (H)	3H (H)	3B (V)	6 (V)	3X (V)		6
7	6065	3840	7 (V)	4V (V)	4A (H)	7 (H)	4D (H)	4 (H)	7
8	6085	3860	8 (H)	4H (H)	4B (V)	8 (V)	4X (V)		8
9	6105	3880	9 (V)	5V (V)	5A (H)	9 (H)	5D (H)	5 (H)	9
10	6125	3900	10 (H)	5H (H)	5B (V)	10 (V)	5X (V)		10
11	6145	3920	11 (V)	6V (V)	6A (H)	11 (H)	6D (H)	6 (H)	11
12	6165	3940	12 (H)	6H (H)	6B (V)	12 (V)	6X (V)		12
13	6185	3960	13 (V)	7V (V)	7A (H)	13 (H)	7D (H)	7 (H)	13
14	6205	3980	14 (H)	7H (H)	7B (V)	14 (V)	7X (V)		14
15	6225	4000	15 (V)	8V (V)	8A (H)	15 (H)	8D (H)	8 (H)	15
16	6245	4020	16 (H)	8H (H)	8B (V)	16 (V)	8X (V)		16
17	6265	4040	17 (V)	9V (V)	9A (H)	17 (H)	9D (H)	9 (H)	17
18	6285	4060	18 (H)	9H (H)	9B (V)	18 (V)	9X (V)		18
19	6305	4080	19 (V)	10V (V)	10A (H)	19 (H)	10D (H)	10 (H)	19
20	6325	4100	20 (H)	10H (H)	10B (V)	20 (V)	10X (V)		20
21	6345	4120	21 (V)	11V (V)	11A (H)	21 (H)	11D (H)	11 (H)	21
22	6365	4140	22 (H)	11H (H)	11B (V)	22 (V)	11X (V)		22
23	6385	4160	23 (V)	12V (V)	12A (H)	23 (H)	12D (H)	12 (H)	23
24	6405	4180	24 (H)	12H (H)	12B (V)	24 (V)	12X (V)		24

Polarization for each transponder denoted in parenthesis.

Courtesy of Westsat Communications

APPENDIX B

YES, I WOULD LIKE TO RECEIVE UP-DATED

INFORMATION ON THE "HIDDEN SIGNALS AND OTHER DATA SERVICES ON THE SATELLITES." (PLEASE SEND SELF-ADDRESSED, STAMPED, NUMBER 10 ENVELOPE WITH THIS FORM)

NAME ______________________________

STREET ______________________________

CITY ____________________ STATE ________________ ZIP __________

MAIL TO: UNIVERSAL ELECTRONICS, INC.
4555 GROVES RD. SUITE 3
COLUMBUS, OHIO, 43232

THIS MATERIAL IS FOR INFORMATION USE ONLY

This work published for the commercial trade to increase awareness of the many non-video opportunities existing in the satellite communications field. The publisher and authors do not approve of the misuse of any of this information in any way by any individuals, companies or groups.